Thomas Göthel

Mechanical Verification of Parameterized Real-Time Systems

Thomas Göthel

Mechanical Verification of Parameterized Real-Time Systems

A Formal Framework and its Application to a Real-Time Operating System Scheduler

Südwestdeutscher Verlag für Hochschulschriften

Impressum / Imprint
Bibliografische Information der Deutschen Nationalbibliothek: Die Deutsche Nationalbibliothek verzeichnet diese Publikation in der Deutschen Nationalbibliografie; detaillierte bibliografische Daten sind im Internet über http://dnb.d-nb.de abrufbar.
Alle in diesem Buch genannten Marken und Produktnamen unterliegen warenzeichen-, marken- oder patentrechtlichem Schutz bzw. sind Warenzeichen oder eingetragene Warenzeichen der jeweiligen Inhaber. Die Wiedergabe von Marken, Produktnamen, Gebrauchsnamen, Handelsnamen, Warenbezeichnungen u.s.w. in diesem Werk berechtigt auch ohne besondere Kennzeichnung nicht zu der Annahme, dass solche Namen im Sinne der Warenzeichen- und Markenschutzgesetzgebung als frei zu betrachten wären und daher von jedermann benutzt werden dürften.

Bibliographic information published by the Deutsche Nationalbibliothek: The Deutsche Nationalbibliothek lists this publication in the Deutsche Nationalbibliografie; detailed bibliographic data are available in the Internet at http://dnb.d-nb.de.
Any brand names and product names mentioned in this book are subject to trademark, brand or patent protection and are trademarks or registered trademarks of their respective holders. The use of brand names, product names, common names, trade names, product descriptions etc. even without a particular marking in this works is in no way to be construed to mean that such names may be regarded as unrestricted in respect of trademark and brand protection legislation and could thus be used by anyone.

Coverbild / Cover image: www.ingimage.com

Verlag / Publisher:
Südwestdeutscher Verlag für Hochschulschriften
ist ein Imprint der / is a trademark of
AV Akademikerverlag GmbH & Co. KG
Heinrich-Böcking-Str. 6-8, 66121 Saarbrücken, Deutschland / Germany
Email: info@svh-verlag.de

Herstellung: siehe letzte Seite /
Printed at: see last page
ISBN: 978-3-8381-3460-4

Zugl. / Approved by: Berlin, TU, Diss., 2012

Copyright © 2012 AV Akademikerverlag GmbH & Co. KG
Alle Rechte vorbehalten. / All rights reserved. Saarbrücken 2012

Abstract

Real-time systems often have to be able to cope with an unbounded number of components. For example, a real-time operating system scheduler manages arbitrarily many threads or a bus system copes with arbitrarily many connected devices. Such systems can be characterized as parameterized systems. The number of (homogeneous) components is the parameter of these systems. This makes their formal verification hard because standard verification techniques for finite system models (for example model checking) can be employed directly only for instances of the system. There exist several approaches for the (automatic) verification of special sub-classes of (mostly untimed) parameterized systems. However, approaches or tools that enable the comprehensive and mechanical verification of parameterized real-time systems with complex system topologies are still missing.

In this thesis, we overcome this problem by providing a framework for the mechanical, comprehensive, and semi-automatic verification of parameterized real-time systems. At its core we employ the process calculus Timed CSP, which is well-suited to describe the functional behavior as well as the non-functional timing behavior of systems. Our main contribution is threefold: First, we have developed a formalization of the operational semantics of Timed CSP together with notions of bisimulation equivalences in the Isabelle/HOL theorem prover. Second, in order to specify (timing) properties of systems, we provide a (mechanized) timed extension of Hennessy-Milner logic. Together with our formalization of Timed CSP and corresponding bisimulations, this enables the comprehensive and mechanical verification of possibly infinite (parameterized) real-time systems in a modular fashion. In particular, by providing both bisimulation and property based verification, we allow the developer to divide the verification problem into subproblems, which eases verification significantly. At the same time, the theorem prover ensures that no corner cases can be overlooked and all proofs are guaranteed to be correct. Finally, we support the development process of such systems with the integration of automatic verification tools. To this end, we have enriched our framework with transformation engines with which (finite) Timed CSP specifications can be transformed to a discrete dialect of CSP and to UPPAAL timed automata. By this, the FDR2 refinement checker and the UPPAAL tool suite can be used to explore and verify finite instances automatically. Thus, possible design flaws can be detected and corrected early in the development cycle and prior to the relatively time-consuming task of interactive theorem proving.

We show the applicability of our framework using the case study of a parameterized real-time operating system scheduler. Thereby, we demonstrate the benefits of the proposed mapping of Timed CSP to automatically analyzable languages. Furthermore, we show the effectiveness of our theorem proving approach for the comprehensive verification of parameterized real-time systems.

The motivation for our work comes from the observation that, due to their increasing complexity, embedded systems are often built atop trusted cores such as bus systems and real-time operating systems. Their correct functioning thus heavily depends on the correctness of the underlying cores. In the area of safety-critical embedded real-time systems, simulation and testing are not sufficient because they cannot ensure the absence of critical errors. Formal verification tackles this problem. To be of practical relevance, however, formal verification techniques should be mechanized and automatized as far as possible. The mechanization comes with the additional benefit of precluding verification faults as they can occur in hand-written proofs. Our case study forms an important core part of many safety-critical systems. In summary, we show how such cores can be mechanically verified and how their development can be supported with the help of automatic verification tools.

Zusammenfassung

Echtzeitsysteme müssen oftmals in der Lage sein, mit beliebig vielen Komponenten umgehen zu können. Der Scheduler eines Echtzeitbetriebssystem, zum Beispiel, verwaltet beliebig viele Threads. Derartige Systeme fallen in die Klasse der parametrisierten Systeme. Deren formale Verifikation ist sehr schwierig, da Standard-Verifikationstechniken für endliche Systemmodelle (z.b. Model Checking) nur für Instanzen des Systems direkt eingesetzt werden können. Es existieren Ansätze für die automatische Verifikationen spezieller Subklassen von parametrisierten Systemen. Bisher fehlt es jedoch an Ansätzen für die umfassende und maschinelle Verifikation von parametrisierten Echtzeitsystemen mit komplexen Systemtopologien.

In dieser Arbeit präsentieren wir ein Framework zur maschinellen, umfassenden und semi-automatischen Verifikation parametrisierter Echtzeitsysteme. Im Kern unseres Frameworks verwenden wir den Prozesskalkül Timed CSP, mit dem sowohl das funktionale Verhalten als auch das nicht-funktionale zeitliche Verhalten von Systemen erfasst werden kann. Unsere wichtigsten Beiträge sind: Erstens, eine Formalisierung der operationalen Semantik von Timed CSP sowie Bisimulationsäquivalenzen im Theorembeweiser Isabelle/HOL. Zweitens stellen wir eine (mechanisierte) zeitbehaftete Erweiterung der Hennessy-Milner Logik zur Verfügung. Zusammen mit der Formalisierung von Timed CSP und den entsprechenden Bisimulationen ermöglicht dies die umfassende und maschinelle Verifikation von möglicherweise unendlichen (parametrisierten) Echtzeitsystemen in modularer Art und Weise. Dadurch dass wir sowohl bisimulations- als auch eigenschaftsbasierte Verifikation ermöglichen, erlauben wir dem Entwickler insbesondere, das Verifikationsproblem in Teilprobleme zu zerlegen. Dies vereinfacht die Verifikation erheblich. Dabei stellt der Theorembeweiser sicher, dass alle Beweise lückenlos und damit garantiert korrekt sind. Unser dritter wesentlicher Beitrag ist die Unterstützung des Entwicklungsprozesses durch automatische Verifikationswerkzeuge. Dazu haben wir unser Rahmenwerk um Transformationswerkzeuge angereichert, mit denen (endliche) Timed CSP Spezifikationen in einen diskreten Dialekt von CSP and in UPPAAL timed automata transformiert werden können. Somit können der Verfeinerungschecker FDR2 und der Model Checker UPPAAL eingesetzt werden, um endliche Systeminstanzen automatisch zu simulieren und zu verifizieren. Dies ermöglicht es, Fehler im Design schon vor dem relativ aufwändigen interaktiven Theorembeweisen aufzudecken.

Die Anwendbarkeit unseres Frameworks zeigen wir anhand der Fallstudie eines Schedulers eines Echtzeitbetriebssystems. Damit demonstrieren wir einerseits den Gewinn durch unsere Transformationen von Timed CSP in automatisch analysierbare Sprachen. Andererseits zeigen wir die Effektivität unseres Theorembeweiser-Ansatzes zur formalen Verifikation parametrisierter Echtzeitsysteme.

Die Motivation dieser Arbeit leitet sich von der Beobachtung ab, dass eingebettete Systeme mit steigender Komplexität oft auf Kernkomponenten wie Bussystemen und Echtzeitbetriebssystemen aufsetzen. Das korrekte Funktionieren dieser Systeme hängt damit stark von der Verlässlichkeit dieser Kernkomponenten ab. In dem Bereich der sicherheitskritischen eingebetteten Echtzeitsystemen reicht es jedoch nicht aus, diese zu simulieren oder zu testen, da damit die Abwesenheit von Fehlern nicht gewährleistet werden kann. Formale Verifikation adressiert dieses Problem. Um jedoch praktisch einsetzbar zu sein, sollten formale Verifikationstechniken mechanisiert und so weit wie möglich automatisiert sein. Die Mechanisierung hat zusätzlich den Vorteil, dass sie Fehler im Prozess der Verifikation ausschließen. Unsere Fallstudie stellt eine wichtige Kernkomponente von sicherheitskritischen Systemen dar. Zusammenfassend zeigen wir, wie derartige Kernkomponenten maschinell verifiziert werden können und wie mithilfe von automatischen Verifikationswerkzeugen ihre Entwicklung unterstützt werden kann.

Contents

1 **Introduction** — 13
  1.1 Problem — 14
  1.2 Objectives — 15
  1.3 Proposed Solution — 16
  1.4 Motivation — 17
  1.5 Main Contributions — 18
  1.6 Context of this Work — 19
  1.7 Outline — 20

2 **Background** — 21
  2.1 Labeled Transition Systems and Correctness of Processes — 22
    2.1.1 Labeled Transition Systems — 23
    2.1.2 Extended Transition Relations — 24
    2.1.3 Bisimulations — 25
    2.1.4 Hennessy-Milner Logic — 28
  2.2 Timed Communicating Sequential Processes — 30
    2.2.1 CSP — 30
    2.2.2 Timed CSP — 40
    2.2.3 The FDR2 Refinement Checker — 46

Contents

- 2.3 UPPAAL Timed Automata 47
  - 2.3.1 Syntax 48
  - 2.3.2 Semantics 49
  - 2.3.3 The UPPAAL Tool Suite 52
- 2.4 The Isabelle Theorem Prover 55
  - 2.4.1 Structure of Isabelle Theories 56
  - 2.4.2 Performing Proofs in Isabelle 57
  - 2.4.3 Datatypes, Functions and (Co)inductive Sets . . . 60
  - 2.4.4 Writing Down Proofs 65
  - 2.4.5 Modular Verification with Locales 66
- 2.5 Summary 70

3 Related Work   71
- 3.1 Formal Verification of Real-Time Systems 72
  - 3.1.1 Model Checking Timed Systems 72
  - 3.1.2 Discretization-based Analysis of Timed CSP . . . 73
  - 3.1.3 Analyzing Timed CSP using Timed Automata Models 74
  - 3.1.4 Further Approaches for the Analysis of Timed CSP   74
  - 3.1.5 Timed Modal Logics 76
- 3.2 Formalization of Process Algebras in Theorem Provers . . 77
  - 3.2.1 IMPS (Interactive Mathematical Proof System) . . 77
  - 3.2.2 PVS (Prototype Verification System) 78
  - 3.2.3 HOL System 79
  - 3.2.4 Isabelle 80
- 3.3 Formal Verification of Parameterized Systems 81
  - 3.3.1 Decidable Subclasses 82
  - 3.3.2 Regular Model Checking 82
  - 3.3.3 Abstraction Techniques 83
- 3.4 Summary 87

Contents

4 Mechanical Verification of Parameterized Real-Time Systems   **89**
- 4.1 Modeling Parameterized Real-Time Systems 93
  - 4.1.1 Parameterized Systems Composed of Parallel Processes 94
  - 4.1.2 Reducing Semantical Complexity of Unbounded Parallel Compositions in Parameterized Systems . 95
- 4.2 Validation and Debugging of System Instances 97
  - 4.2.1 Transformation to Timed Automata 98
  - 4.2.2 Transformation to Tock CSP 99
- 4.3 Bisimulation-Based Verification 101
- 4.4 Logic-Based Verification 104
  - 4.4.1 Timed Hennessy-Milner Logic 104
  - 4.4.2 Examples 106
  - 4.4.3 Coinductive Invariants 108
  - 4.4.4 Logical Verification 108
- 4.5 Summary 111

5 Formalization of Timed CSP in the Isabelle/HOL Theorem Prover   **113**
- 5.1 Fundamental Theories 114
  - 5.1.1 Labeled Transition Systems 115
  - 5.1.2 Abstract Bisimulations 116
  - 5.1.3 Timed Hennessy-Milner Logic 119
  - 5.1.4 Preservation of Timed HML under Bisimulation . 121
- 5.2 Formalization of Timed CSP 123
  - 5.2.1 Syntax 123
  - 5.2.2 Operational Semantics 126
  - 5.2.3 Timed CSP as a Timed Labeled Transition System 132
  - 5.2.4 Bisimulation as an Observational Congruence . . . 133
  - 5.2.5 Outlook: Denotational Semantics 134

Contents

    5.3   Coping with Parameterized Systems 137
            5.3.1   Arbitrarily Large Networks of Timed Processes . . 138
            5.3.2   Expressing Properties of Parameterized Real-Time
                      Systems 139
    5.4   Summary 140

6   Integration of Automatic Verification Tools   143
    6.1   Transformation from Timed CSP to Timed Automata . . . 146
            6.1.1   Assumptions 147
            6.1.2   Transformation Rules 148
            6.1.3   Simulation and Automatic Verification in UPPAAL 156
            6.1.4   Discussion 160
    6.2   Transformation from Timed CSP to Tock CSP 161
            6.2.1   Assumptions 162
            6.2.2   Transformation Rules 163
            6.2.3   Automatic Verification with FDR2 170
            6.2.4   Discussion 172
    6.3   Summary 175

7   Case Study - A Real-Time Operating System Scheduler   177
    7.1   Timed CSP Model of a Real-Time Scheduler 178
            7.1.1   Formalizing the Model 179
            7.1.2   Formalizing the Requirements 185
    7.2   Instance Verification using UPPAAL and FDR2 187
            7.2.1   Model Checking in UPPAAL 188
            7.2.2   Refinement Checking in FDR2 190
            7.2.3   Transformation Times 190
    7.3   Comprehensive Verification using Isabelle/HOL 192
            7.3.1   Bisimulation Proof 192
            7.3.2   Verification of the Timed HML Properties 195
    7.4   Summary 198

8 Conclusion and Future Work · 201
    8.1   Conclusion 201
    8.2   Discussion 205
    8.3   Future Work 208

List of Figures   213

List of Definitions   215

Bibliography   217

Contents

	Brief Outline of this Thesis
Chapter 1 Introduction	• Introduction to the field and problem description • Definition of objectives to evaluate our solution • Short description of our solution • Motivation of our work • Summary of the main contributions of this work
Chapter 2 Background	• Labeled transition systems, bisimulations, and Hennessy-Milner logic • CSP, Timed CSP, and the FDR2 refinement checker • Timed automata and the UPPAAL tool suite • The Isabelle/HOL theorem prover
Chapter 3 Related Work	• Discussion of related work based on the objectives given in the introduction

Chapter 4
A Framework for the Mechanical Verification of Parameterized Real-Time Systems
	• Overview of our framework for the mechanical verification of parameterized systems
	    – Description of parameterized real-time systems in Timed CSP
	    – Use of automatic verification tools
	    – Verification using bisimulations
	    – Definition of our timed extension of Hennessy-Milner logic and its use for verification

Chapter 5 Formalization of Timed CSP in the Isabelle/HOL Theorem Prover	• Mechanization of (timed) labeled transition systems, bisimulations, and our timed extension of Hennessy-Milner logic • Mechanization of the Operational Semantics of Timed CSP • Support for Parameterized Real-Time Systems

Contents

Chapter 6 Integration of Automatic Verification Tools	• Transformation from Timed CSP to timed automata and verification of translated models using UPPAAL • Transformation from Timed CSP to tock CSP and verification of translated models using FDR2
Chapter 7 Case Study - A Real-Time Operating System Scheduler	• Application of the entire framework to a real-time scheduling system
Chapter 8 Conclusion and Future Work	• Summary of the main contributions of this thesis • Evaluation according to the objectives given in the introduction • Possible directions for future work

1 Introduction

Embedded systems are often employed in safety-critical areas and their complexity is steadily increasing. Therefore, formal verification gains more and more importance in this area. Unlike testing, formal verification can ensure the absence of design errors, which is crucial in safety-critical systems. Furthermore, if the verification flow is machine-assisted, errors that typically occur in "paper and pencil" proofs are prevented. In this thesis, we consider the formal verification of embedded systems that have to cope not only with a fixed number of components but with an arbitrarily large number of them. Examples of these systems are real-time operating system schedulers that have to be able to manage arbitrarily many threads or bus systems that have to cope with arbitrarily many connected devices. These kinds of systems fall into the class of parameterized systems, i.e., the number of components is a parameter. To verify such systems, all possible instances of the system have to be considered. This task is still an open problem. A particular challenge is to develop a verification approach, which can cope with a large class of system topologies, general kinds of parameterization, and real-time constraints. At the same time, the approach should enable continuous machine assistance and the integration of automated verification support.

Introduction

1.1 Problem

This thesis addresses the problem of mechanically verifying parameterized real-time systems and of supporting their development. A parameterized system is composed of an arbitrarily large number of homogeneous processes. The number of processes in such a system can be interpreted as its parameter. In addition, the network of processes can be controlled by (a) distinguished control process(es). Both, the control process and the network processes may depend on the network size. The typical structure of a parameterized system is therefore $N_n \stackrel{def}{=} C_n \otimes_0 (P_{1,n} \otimes P_{2,n} \otimes \cdots \otimes P_{n,n})$ where C_n is the control process, $P_{i,n}$ is one of the n homogeneous components with index i and $\otimes_0$, $\otimes$ are some kind of parallel composition operators. Thus, parameterized systems are infinite systems because there exist infinitely many instances. The verification problem is the question whether for all n, N_n meets a certain specification. In general, this problem is undecidable. Therefore, automatic verification tools can only be applied to some instances of the parameterized system or special subclasses of parameterized systems have to be considered. The problem with (comprehensive) automatic verification approaches for parameterized systems is that the description of parameterized systems is often rather unnatural and that they are only applicable to relatively restricted classes of parameterized systems. In particular, the considered systems in the literature often form a linear network but not a centralized or star-shaped one. Another problem is that most of these approaches do not cope with real-time models. Finally, the existing approaches do not come with formal comprehensive machine assistance. Thus, the verification itself is not ensured to be correct.

1.2 Objectives

In this thesis, we aim at developing a framework for the mechanical verification of parameterized real-time systems. The framework should fulfill the following criteria:

1. **Coping with general parameterized systems**

   Most approaches for the verification of parameterized systems only cope with very restricted classes of systems. Often, only networks that are organized in a linear or ring-like structure are considered. Our approach should deal with a broad range of parameterized systems, especially with centralized systems like schedulers or bus systems. Here, distinguished control processes are allowed that control the network of components and interact with it. Both, the control process and the network processes, should be allowed to depend on the network size.

2. **Coping with real-time specifications**

   Our framework shall especially enable the convenient verification of real-time specifications. Furthermore, the user shall not be restricted in advance to discrete-time system models as these models possibly abstract from the real world in an unsafe way.

3. **Comprehensive machine assistance**

   Our framework shall be based on verification tools that enable all proofs to be mechanically checked in an interactive theorem prover. This ensures that erroneous "paper and pencil" proofs are prevented.

Introduction

4. **Supporting the development process with automatic (verification) tools**

    When developing the system model, design errors should be detected as early as possible with a high degree of automatization. This would implicitly reduce the relatively expensive process of interactive theorem proving since proofs need not to be started over again too often. Therefore, our framework shall provide possibilities to detect errors prior to the mechanical verification phase.

1.3 Proposed Solution

We propose a framework for the verification and development of parameterized real-time systems. At its core, we use a Timed CSP formalization in the Isabelle/HOL theorem prover. We have formalized the operational semantics of Timed CSP together with different notions of bisimulation equivalence. In order to express and verify (timing) properties of real-time systems, we provide a (mechanized) timed extension of Hennessy-Milner logic. Additionally, we have formalized special support for the description and verification of parameterized systems, which enables us to conveniently perform bisimulation- and property-based proofs about such systems. To support early bug detection during the development of parameterized real-time systems, we have integrated techniques for the transformation of finite Timed CSP processes to automatically analyzable formal languages. We have adapted, extended, and implemented the transformation rules of [Oua01] and of [DHQ$^+$08]. These transformations map Timed CSP processes to tock CSP and timed automata, respectively. These transformed models can then be simulated and automatically verified using the FDR2 refinement checker [GRA05] and the UPPAAL tool suite [BY04]. This orthogonal layer of our approach allows for early detection of design flaws. This implicitly reduces the complexity of performing interactive theorem proving because most of the design flaws can be detected by the prior use of automatic simulation and verification tools.

In summary, our approach has three major advantages: First, we employ Timed CSP for system specification, which allows a large class of parameterized real-time systems to be described and verified. Second, our approach involves comprehensive machine assistance in the Isabelle/HOL theorem prover supporting the mechanical modeling and verification of parameterized real-time systems. Third, we enable the automatic transformation of Timed CSP to other languages allowing for the automatic simulation and verification of (finite) instances of the parameterized system specification.

1.4 Motivation

Embedded systems software is ubiquitous. It occurs in cars, in mobile phones, in power plants and many other areas. While it is not too problematic if the software of a mobile phone contains errors, it is if errors occur in safety-critical areas like cars, planes or nuclear power plants. This can cause high costs or in the worst case the loss of human lives. Due to the increasing complexity of such systems, they are built atop trusted cores like real-time schedulers and bus systems. This means that verifying these cores is an inevitable prerequisite to ensure the correctness of systems depending on these cores. To assure the quality of embedded systems, nowadays mainly test methods are used. They are good in detecting errors in systems but it is not possible to show their absence. To show that no error can occur in a safety-critical system, even not in corner cases, formal methods should be employed.

Introduction

Among others, three highly important properties of embedded systems distinguish them from usual computer systems: The interaction with the environment, the requirement of timeliness, and the high number of concurrent tasks. Therefore, we consider Timed CSP as a suitable formalism to cope with embedded real-time systems. Timed CSP is a timed extension of the process calculus *Communicating Sequential Processes (CSP)*. It comes with formal semantics that allows for reasoning about reactive, concurrent and timed systems.

1.5 Main Contributions

In summary, the contributions of this thesis are the following

- Formalization of Timed CSP in the Isabelle/HOL theorem prover to support the convenient specification and verification of parameterized real-time systems.

- Mechanization of bisimulation-based and property-based verification techniques in Isabelle/HOL that enable the modular verification by dividing the verification problems into subproblems.

- Extension and implementation of transformation engines from Timed CSP processes to UPPAAL timed automata, which can be automatically simulated and verified using the UPPAAL tool suite.

- Extension and implementation of transformation engines from Timed CSP processes to discrete tock CSP models, which can be automatically verified using the FDR2 refinement checker.

- Definition of a development and verification framework for general parameterized real-time systems integrating interactive theorem proving, simulation and model checking.

- Application of our framework to the case study of a real-time operating system scheduler.

1.6 Context of this Work

The context of this work is the DFG-funded VATES[1] project [GHJ07]. Its aim is to develop techniques that support the construction and the formal machine-assisted verification of safety-critical embedded real-time systems. Thereby, the whole development chain is considered beginning with an abstract specification down to finally executable code. The key idea in the VATES project is to show that an implementation of a safety-critical real-time system, given as executable LLVM [LA04] intermediate code, conforms to a Timed CSP specification. The assumed development flow is that a designer develops a Timed CSP specification and mechanically verifies it with respect to crucial (timing) properties. Then, a developer implements the Timed CSP specification in a high-level programming language like C++ and compiles it to LLVM. From the LLVM code, we extract a low-level CSP model [KBG⁺11], which is shown to refine the abstract specification. With that, we achieve the formal conformance between the abstract specification and its LLVM implementation given the correctness of the extraction from LLVM to low-level CSP. To also mechanically verify that this extraction is correct (which is ongoing work), we use our formalization of Timed CSP [GG10a] and developed a formalization of a timed operational semantics of LLVM in Isabelle/HOL [Bar11]. We presented the overall VATES approach in [GBGK10, BGG10]. The work described in this thesis supports the first phase in the VATES approach, where an abstract Timed CSP model is mechanically shown to fulfill crucial (timing) properties.

---

[1] <u>V</u>erification <u>a</u>nd <u>T</u>ransformation of <u>E</u>mbedded <u>S</u>ystems

Introduction

1.7 Outline

The rest of this thesis is organized as follows: In Chapter 2, we describe the necessary background of this thesis. We introduce the notions of (timed) labeled transition systems, bisimulations, and Hennessy-Milner logic. We then we briefly introduce the (Timed) CSP process algebra focusing on its operational semantics. Furthermore, we describe the simulation/verification tools employed in this thesis: the FDR2 refinement checker, the UPPAAL tool suite, and the Isabelle/HOL theorem prover. In Chapter 3, we discuss related work of this thesis. In Chapter 4, we give an overview of our framework for the mechanized verification of parameterized real-time systems. In Chapter 5, we present our formalization of timed labeled transition systems, our timed extension of Hennessy-Milner logic, and of Timed CSP in the Isabelle/HOL theorem prover and present special support needed to describe and mechanically verify parameterized real-time systems. In Chapter 6, we present our extensions of transformation rules from Timed CSP to timed automata and to tock CSP, respectively. Furthermore, we describe how transformed models can be automatically analyzed using UPPAAL and FDR2 for early detection of design flaws in parameterized real-time specifications. A real-time operating system scheduler is used as a case study. We present the results of the application of our framework to it in Chapter 7. We close this thesis with a conclusion and possible directions for future work in Chapter 8.

2 Background

In this chapter, we introduce the necessary background information concerning mathematical languages and models that we use in this thesis and briefly describe the tools to support the mechanical and automatized verification of corresponding models.

We begin with the introduction of the mathematical notions of (timed) labeled transition systems (LTSs), different kinds of bisimulation [Mil89] for expressing equivalences of processes, and (untimed) Hennessy-Milner logic (HML) [HM80] for property-oriented specifications of processes in Section 2.1. In Section 2.2, we introduce the process algebra CSP [Hoa85] and present its timed extension Timed CSP [Sch99]. We especially focus on the operational semantics as this is the kind of semantics our formal verification framework is based on. Since Timed CSP can be interpreted as a timed LTS based on its operational semantics, we can instantiate the notions of bisimulation and HML in this context. Furthermore, we briefly introduce the FDR2 refinement checker [GRA05]. FDR2 is based on the denotational semantics of CSP and enables the verification of untimed CSP processes (and Timed CSP processes to a certain degree) by means of refinement. Another established language for modeling and verification of timed systems is that of timed automata [AD94]. We introduce this formalism in Section 2.3, focusing on the dialect that the UP-

Background

PAAL tool suite [BY04] is based on. This dialect enables the application of abstraction techniques such that automatic verification tools, for example the UPPAAL model checker, can be applied. We also briefly describe the capabilities of the UPPAAL tool suite. A brief introduction to Higher-Order-Logic (HOL) and its mechanization in the Isabelle/HOL theorem prover [NPW02] is given in Section 2.4. Additionally, we present important capabilities of Isabelle that are very useful for our formalizations of timed LTSs and Timed CSP in Isabelle/HOL. We close this chapter with a summary in Section 2.5.

2.1 Labeled Transition Systems and Correctness of Processes

The notion of LTSs gives a very general formalism, which is widely used to describe the behavior of systems. An LTS consists of states and labeled edges between them. The states are also often called processes because they describe how the system can evolve subsequently. To verify the correctness of processes, there exist two general approaches: equivalence verification and property verification. Equivalence verification means that behavioral equivalence with another process, which serves as a specification, is used as the correctness criterion. Property verification, on the other hand, means that correctness criteria are expressed by one or more logical properties, for example, safety, liveness, or timing properties. To show the behavioral equivalence between processes, bisimulations can be used, which offer a convenient way to express and verify equivalences. Based on the underlying LTS, different kinds of bisimulations can be established for different levels of abstraction. A convenient way for expressing properties of processes is HML. It is a modal logic specifying allowed executions of a process. The connection between bisimulation and HML is that bisimi-

2.1 Labeled Transition Systems and Correctness of Processes

lar processes satisfy the same logical properties. Under certain conditions the other direction holds as well, i.e., if two processes satisfy exactly the same set of properties, they are bisimilar. In the following, we introduce the notion of (timed) LTS, important kinds of bisimulation, and the basic concepts of HML.

2.1.1 Labeled Transition Systems

In this thesis, we consider two types of transition systems: LTSs and timed LTSs. In contrast to "simple" LTSs, timed LTSs additionally consider a time domain and require further properties of the transition relation with respect to time.

Definition 1 (Labeled Transition System) *An LTS is given by a tuple (S,T,A), where S is a set of states, $T \subseteq (S \times A \times S)$ is a labeled transition relation, and A is a label set.*

In the context of an LTS, we also call a state $P \in S$ a *process*. For $(P, \alpha, P') \in T$, we write $P \xrightarrow{\alpha} P'$. If $P \xrightarrow{\alpha} P'$ we call P' an α-derivative of P. Typically, there is a distinguished event $\tau \in A$, which is interpreted as an internal event, i.e., it is not visible to a given environment. In the following, we assume τ to be an element of A.

Definition 2 (Timed Labeled Transition System) *A timed LTS is given by a tuple (S,T,A,D), where S is a set of states, $T \subseteq S \times (A \cup D) \times S$ is the timed labeled transition relation, A is a label set and D is a time domain (such as $\mathbb{N}$ or $\mathbb{R}_{\geq 0}$) with A and D disjoint. We also write $P \stackrel{d}{\rightsquigarrow} P'$ for $(P,d,P') \in T$ with $d \in D$. For each timed LTS, we require the following two properties to be fulfilled.*

- $\forall t_1\ t_2.\ P \stackrel{t_1+t_2}{\rightsquigarrow} Q \longrightarrow \exists P'.\ P \stackrel{t_1}{\rightsquigarrow} P' \wedge P' \stackrel{t_2}{\rightsquigarrow} Q$
- $\forall t.\ (P \stackrel{t}{\rightsquigarrow} P' \wedge P \stackrel{t}{\rightsquigarrow} P'') \longrightarrow P' = P''$

Background

The first property states that compound timed steps can be split into consecutive timed steps. The second property states that the timed LTS is time deterministic.

Note that every timed LTS (S,T,A,D) can also be interpreted as a "simple" LTS $(S,T,A \cup D)$ by joining the label set with the time domain.

2.1.2 Extended Transition Relations

The transition relation of a (timed) LTS is used to describe the behavior of processes. To be able to define different kinds of bisimulations, we first define extended transition relations that abstract away from internal behavior of processes in various degrees.

Given an LTS (S,T,A), then we define a (weak) extended transition relation $\longrightarrow_w \subseteq S \times A \times S$ to abstract from internal steps as follows[1].

Definition 3 (Weak Extended Transition Relation)

1. If $P \xrightarrow{\tau}^* P'$, then $P \xrightarrow{\tau}_w P'$.

$$(\text{i.e., } P \xrightarrow{\tau} \ldots \xrightarrow{\tau} P')$$

2. If $P \xrightarrow{\tau}^* P_1$, $P_1 \xrightarrow{a} P_2$, and $P_2 \xrightarrow{\tau}^* P'$ with $a \neq \tau$, then $P \xrightarrow{a}_w P'$.

$$(\text{i.e., } P \xrightarrow{\tau} \ldots \xrightarrow{\tau} P_1 \xrightarrow{a} P_2 \xrightarrow{\tau} \ldots \xrightarrow{\tau} P')$$

For a timed LTS (S,T,A,D), we define a (weak timed) extended transition relation $\longrightarrow_{wt} \subseteq S \times (A \cup D) \times S$ that additionally allows for the aggregation of consecutive (weak) timed steps with the help of $\longrightarrow_w$ (defined over $A \cup D$).

---

[1] $\xrightarrow{\tau}^*$ denotes the reflexive-transitive closure of the original transition system with respect to τ.

2.1 Labeled Transition Systems and Correctness of Processes

Definition 4 (Weak Timed Extended Transition Relation)

1. If $P \xrightarrow{\alpha}_w P'$ and $\alpha \in A$, then $P \xrightarrow{\alpha}_{wt} P'$.

2. If for all $i \in \{0, \ldots, n\}$ (for some arbitrary $n \in \mathbb{N}$) $P_i \xrightarrow{t_i}_w P_{i+1}$ with $t_i \in D$ and $\sum_{i=0}^{n} t_i = t$, then $P_0 \xrightarrow{t}_{wt} P_{n+1}$.

$$(i.e.\ P_0 \xrightarrow{\tau}^* \bullet \xrightarrow{t_0} \bullet \xrightarrow{\tau}^* \bullet$$

$$\ldots$$

$$\bullet \xrightarrow{\tau}^* \bullet \xrightarrow{t_n} \bullet \xrightarrow{\tau}^* P_{n+1})$$

Furthermore, we derive a transition relation that is used to describe that a process can perform a timed event (t, a). This definition is used in particular to define our timed extension of HML in Section 4.4.

Definition 5 (Time-Event Step) *If $t \in D$ and $a \in A$ with $a \neq \tau$ is some visible event, we define time-event steps as*

$$P \xRightarrow{(t,a)} Q \overset{def}{=} \exists P'.\ P \xrightarrow{t}_{wt} P' \wedge P' \xrightarrow{a}_{wt} Q$$

This means that P can communicate the event a after idling (and possibly performing internal steps) for t time units.

2.1.3 Bisimulations

Bisimulations, as for example introduced in [Mil89], are a strong and convenient proof principle for showing semantical equivalence of processes (i.e., states of a (timed) LTS). Based on the level of abstraction, different kinds of bisimulations can be introduced. In this subsection, we consider strong, weak and weak timed bisimulation. While strong and weak bisimulation can be established on "simple" LTS as they do not consider time explicitly, weak timed bisimulation explicitly considers timed steps.

Background

Strong Bisimulation The idea of strong bisimulation is to define two processes equivalent if they can always match each of their transition steps adequately. Formally, this is expressed in the following definition.

Definition 6 (Strong Bisimulation) *A relation $R \subseteq S \times S$ is called a strong bisimulation on an LTS (S,T,A) if the following properties hold. For all $(P,Q) \in R$ and $\alpha \in A$:*

1. *If $P \xrightarrow{\alpha} P'$, then there is a Q' with $Q \xrightarrow{\alpha} Q'$ and $(P',Q') \in R$.*
2. *If $Q \xrightarrow{\alpha} Q'$, then there is a P' with $P \xrightarrow{\alpha} P'$ and $(P',Q') \in R$.*

This means that for each step of P a corresponding step must exist for process Q such that the reached processes are again in the bisimulation relation and vice versa. By the quantification over all pairs in the bisimulation relation, this "game" can be continued infinitely.

Weak Bisimulation In the case where (S,T,A) is an LTS and $\tau \in A$ is a distinguished internal event, weak bisimulation can be defined that abstracts away from the internal behavior of processes to a certain degree.

Definition 7 (Weak Bisimulation) *A relation $R \subseteq S \times S$ is called a weak bisimulation on an LTS (S,T,A) if the following property holds. For all $(P,Q) \in R$ and $\alpha \in A$*

1. *If $P \xrightarrow{\alpha} P'$, then there is a Q' with $Q \xrightarrow{\alpha}_w Q'$ and $(P',Q') \in R$.*
2. *If $Q \xrightarrow{\alpha} Q'$, then there is a P' with $P \xrightarrow{\alpha}_w P'$ and $(P',Q') \in R$.*

Note that in this definition α can also be $\tau \in A$, which implies that a τ step has to be matched by zero or more consecutive τ steps. In the case of a step labeled with $\alpha \neq \tau$, it has to be matched with a compound step consisting of an α step with arbitrarily many internal steps before and after it.

2.1 Labeled Transition Systems and Correctness of Processes

Weak Timed Bisimulation In the context of timed LTSs of the form (S,T,A,D), we define weak timed bisimulations (like, for example, in [LY93]). Weak timed bisimulation again abstracts away from internal behavior. Furthermore, a single timed step may be answered by arbitrarily many consecutive timed steps, where the summed duration is equal to the original time span. Furthermore, arbitrarily many internal steps are allowed between these single timed steps.

Definition 8 (Weak Timed Bisimulation) *A relation $R \subseteq S \times S$ is called a weak timed bisimulation on a timed LTS (S,T,A,D) if the following property holds. For all $(P,Q) \in R$ and $\beta \in A \cup D$:*

1. *If $P \xrightarrow{\beta} P'$, then there is a Q' with $Q \xrightarrow{\beta}_{wt} Q'$ and $(P',Q') \in R$.*
2. *If $Q \xrightarrow{\beta} Q'$, then there is a P' with $P \xrightarrow{\beta}_{wt} P'$ and $(P',Q') \in R$.*

For every kind of bisimulation we call P and Q (strong, weak, or weak timed) bisimilar if there exists a corresponding bisimulation R such that $(P,Q) \in R$. If R and R' are two bisimulations, then the union $R \cup R'$ is also a bisimulation. The generalization is the union of all bisimulations: $\hat{R} \stackrel{def}{=} \bigcup \{R \mid R \text{ is a bisimulation}\}$. This is actually the greatest bisimulation, leading to the fact that two processes P and Q are bisimilar if and only if $(P,Q) \in \hat{R}$. All considered kinds of bisimulation allow us to identify semantically equivalent processes with respect to a (timed) LTS. Note that they all have the same structure, i.e., for two equivalent processes P and Q, every simple step of process P must be answered by a possibly "complex" step of Q, and vice versa. The various complex steps are used for abstracting from details of single steps to a certain degree.

Bisimulations are well-suited for proving the equivalence of processes. However, it tends to be inconvenient to describe properties in terms of processes and to prove the property process bisimilar to the implementation process. To additionally be able to express properties of single processes,

Background

we employ a timed extension of HML in our framework. The original version of this logic for (untimed) LTSs is covered in the following subsection.

2.1.4 Hennessy-Milner Logic

HML was introduced in [HM80]. It is a small modal logic that is used to describe properties of processes in an LTS (S, T, A). Its basic modality operator is called *possibility*. It asserts for a process P that there exists an α-derivative[2] of P satisfying a certain formula. Using negation, a *necessity* operator can be introduced asserting for P that all α-derivatives of P satisfy a certain formula. The logic gains its expressive power by the arbitrary interleaving of these modalities. HML was developed to logically characterize strong bisimulations, i.e., two processes are strong bisimilar iff they satisfy exactly the same formulae. This holds for the case that the underlying LTS is image-finite, i.e., for every process P and label α there are only finitely many α-derivatives of P. In [Mil89], the logic was extended by allowing infinite conjunctions in formulae. Then, the characterization theorem for strong bisimulation can also be established in the case that the underlying LTS is not image-finite. Furthermore, the characterization for weak bisimulations is also established by replacing the possibility operator with a weaker version.

The syntax of HML with respect to an LTS (S, T, A) where α ranges over the label set A is given as follows.

Definition 9 (Syntax of Hennessy-Milner Logic)

$$\phi := tt \mid \neg \phi \mid \phi_1 \wedge \phi_2 \mid \langle \alpha \rangle \phi$$

We take *Form* to be the set of all formulae ϕ. The logic consists of the usual logical operators (*true* (tt), *negation* ($\neg$) and *conjunction* ($\wedge$)), and a

---
[2]Remember that if $P \xrightarrow{\alpha} P'$ then P' is called α-derivative of P.

2.1 Labeled Transition Systems and Correctness of Processes

modality operator $\langle \_ \rangle\_$ called *possibility*. The intuition of the possibility operator is that a process P satisfies $\langle \alpha \rangle \phi$ if there is an α-derivative of P for which ϕ holds.

Definition 10 (Semantics of Hennessy-Milner Logic) *The semantics of fulfillment with respect to an LTS (S,T,A) is given by $\models \,\subseteq (S \times \text{Form})$.*

$$P \models tt \quad \text{for all } P$$
$$P \models \neg \phi \quad \text{iff} \quad \neg(P \models \phi)$$
$$P \models \phi_1 \wedge \phi_2 \quad \text{iff} \quad P \models \phi_1 \wedge P \models \phi_2$$
$$P \models \langle \alpha \rangle \phi \quad \text{iff} \quad \exists Q.\, P \xrightarrow{\alpha} Q \wedge Q \models \phi$$

The usual abbreviations for $f\!f, \vee, \longrightarrow, \ldots$ can be defined straightforwardly. A further important abbreviation is the *necessity* operator $[\alpha]\phi$, which is defined as $\neg \langle \alpha \rangle \neg \phi$. Some process P satisfies formula $[\alpha]\phi$ if all α-derivatives of P satisfy formula ϕ. Formally, this is expressed as $P \models [\alpha]\phi$ iff $\forall Q.$ if $P \xrightarrow{\alpha} Q$ then $Q \models \phi$.

To exemplify the expressiveness of HML, consider the following formulae.

1. $P \models \langle a \rangle tt$ – it is possible to carry out an a experiment on p
2. $P \models [a]f\!f$ – p is a-deadlocked
3. $P \models \langle a \rangle ([b]f\!f \vee [c]f\!f)$ – it is possible via an a experiment to get into a state that is either b-deadlocked or c-deadlocked
4. $P \models [a](\langle b \rangle [c]f\!f)$ – at the end of any a experiment a b experiment is possible, which will leave the program in a state that is c deadlocked.

In this subsection, we have briefly introduced the basic (untimed) version of HML. We extend this logic in Section 4.4 by replacing the basic

Background

modality operator with a weak timed version of it in order to express timing properties of processes in timed LTS. We especially consider the timed LTS of Timed CSP, given by its timed operational semantics, which is introduced in the following section.

2.2 Timed Communicating Sequential Processes

Timed Communicating Sequential Processes (Timed CSP) is a process calculus, which extends CSP with timed process operators and timed semantics. In the following, we introduce the formalism of CSP and then proceed with the description of its extensions to Timed CSP. Finally, we give a brief introduction of the FDR2 refinement checker, which is based on untimed semantics of CSP but can be employed for the verification of Timed CSP processes to a certain degree.

2.2.1 CSP

The process calculus CSP (Communicating Sequential Processes) was developed in the late 1970s and was largely stable by the mid 1980s [Hoa85]. It is capable of specifying and verifying reactive and concurrent systems. The modeling of communication between concurrent processes plays a key role in CSP. It is equipped with a rich set of process operators for defining possibly infinite transition systems. Its *processes* perform *events* that are both atomic and instantaneous. The set of events that may be communicated by a process builds its alphabet Σ. If a process offers an event with which its environment agrees to synchronize, the event is performed. Additionally, a process can possibly engage in the τ and $\sqrt{}$ events. These distinguished events are not part of the process alphabet Σ. The internal

2.2 Timed Communicating Sequential Processes

event τ is performed by a process when some internal transition happens that the environment cannot influence. The terminating event $\sqrt{}$ is communicated when a process successfully terminates. The formal semantics of CSP can be given in terms of denotational semantics or in terms of the operational semantics. In the denotational semantics, a process is mapped to a mathematical domain that abstractly describes its behavior in terms of traces and refusals. In the operational semantics, semantical deduction rules describe how a process can evolve step-wise. The latter enables that a process can be thought of as a (possibly infinite) transition system.

Syntax

CSP processes are used to describe the communication of events in a concise way. Let Σ be an alphabet and let $\tau \notin \Sigma$ and $\sqrt{} \notin \Sigma$ be two special events not included in the alphabet Σ. Furthermore, there is a set $\mathcal{V}$ of process variables to describe recursive and infinite processes as described below. An excerpt of the syntax of CSP is given by

Definition 11 (Syntax of CSP)

$$P := STOP \mid SKIP \mid a \to P \mid x : A \to P_x \mid P;P \mid$$
$$P \square P \mid P \triangle P \mid P \sqcap P \mid P \parallel_A P \mid P \backslash A \mid$$
$$X \mid \ldots$$

In this context, a ranges over Σ denoting a single event, A ranges over Σ and denotes a set of events and X ranges over $\mathcal{V}$ denoting a process variable. We take *CSP* to be the set of all CSP processes.

The process *STOP* cannot do anything except to deadlock. *SKIP* cannot do anything except to terminate indicated by the communication of the special event $\sqrt{}$. The *Prefix* process $a \to P$ first communicates the event a and then behaves like process P. The *Menu Choice* process $a : A \to P_a$

Background

generalizes this concept in that it can communicate an event out of A and then behaves like P_a. In CSP, events are often structured. For example, channels of the form "channel $a : \{1,2,3\}.\{1,2,3\}$" can be defined in the FDR2 refinement checker. This is essentially a shortcut for introducing events $a.1.1, a.1.2$ and so on. The special syntax $a?x?y \to P$ can be used to conveniently work with structured channels. For example, the process $a?x?y \to b!(x*y) \to STOP)$ receives two input values on channel a and outputs their product on channel b. The construction $c?x \to P_x$ can, be defined using the *Menu Choice* operator: $e : \{c.x.\ x \in \{1,2,3\}\} \to P_{value(e)}$. Therefore, no new syntactical operators are needed to support this common feature of CSP. *Sequential Composition* of processes is written as $P;Q$ and behaves like P as long as it has not terminated, then the process behaves like Q. There are two important operators for choice: *Internal Choice* ($\sqcap$) and *External Choice* ($\square$). *Internal Choice* is resolved nondeterministically whereas *External Choice* is resolved by the environment deciding for the one or the other process. The *Interrupt* construction ($\triangle$) allows the second process to interrupt the first one on visible events as long as the first process has not terminated. *Parallel Composition* is written as $P \parallel_A Q$, describing the composition of concurrent processes P and Q that must synchronize on all events in A but can act independently for all other events. Abstraction is realized by the *Hiding* operator ($P \setminus A$), which means that all events in A are hidden in P, i.e., they become internal τ events. This implies that the environment cannot control the communication of the hidden events as they are resolved internally. The *Process Variables* are used to model (mutually) recursive processes using a mapping $asg :: X \Rightarrow CSP$ called process variable assignment. The semantics of process variables is given by a process variable assignment and the semantic rule for process variables as described below.

As an example, the recursive process $P = a \to P$ is formalized using a process variable P, which is mapped to $a \to P$, i.e., $asg(P) = a \to P$.

2.2 Timed Communicating Sequential Processes

In the CSP literature, recursion is also sometimes introduced by an explicit recursion operator. Then, the process P above would be defined as $P \stackrel{def}{=} \mu X \bullet a \rightarrow X$.

Infinite processes can be described using an infinite set of process variables. Let, for example, $\{C_i \,.\, i \in \mathbb{N}\} \subseteq \mathcal{V}$. Then, a simple counter process C can be realized by

$asg(C_i) = $ **if** $i = 0$ **then** $succ \rightarrow C_1$
$\qquad\qquad\qquad$ **else** $(pred \rightarrow C_{i-1}) \square (succ \rightarrow C_{i+1})$

and $C \stackrel{def}{=} C_0$.

For convenience, we additionally define an abbreviated guard operator

$g \,\&\, P \stackrel{def}{=}$ **if** g **then** P **else** $STOP$

Using this operator, we can define the process variable assignment for the counter process above in an even more convenient way as

$asg(C_i) = succ \rightarrow C_{i+1}$
$\qquad\qquad\square\, i > 0 \,\&\, pred \rightarrow C_{i-1}$

The semantics of CSP can be defined as a denotational and as an operational semantics. The common verification tools for CSP such as the FDR2 refinement checker are based on the denotational semantics. For giving a more intuitive meaning to CSP processes and to be able to employ coinductive verification techniques such as bisimulation-based reasoning, the operational semantics can be used. In the following, both semantics are briefly introduced. However, as we focus on the operational semantics of (Timed) CSP in this thesis, we give an operational characterization of the denotational semantics (similar to [Sch99]) instead of directly introducing the denotational semantics.

Background

Operational Semantics

The processes of CSP can be interpreted by means of an LTS. The LTS defined by the operational semantics is $(CSP, \longrightarrow, \Sigma \cup \{\tau, \sqrt{}\})$ with $\longrightarrow$ defined as below. It gives an intuitive understanding of the behavior of processes and additionally gives the opportunity to instantiate coalgebraic notions like bisimulation in the context of CSP. In the following, we introduce the semantical rules for each of the process terms of CSP.

SKIP The *SKIP* process can successfully terminate indicated by the communication of $\sqrt{}$ and then behaves like *STOP*, which cannot do anything.

$$\overline{SKIP \xrightarrow{\sqrt{}} STOP}$$

Prefix The *Prefix* process can communicate the event a and then behaves like P.

$$\overline{(a \to P) \xrightarrow{a} P}$$

Menu Choice The generalization of *Prefix* is *Menu Choice* and allows for the communication of an event out of a given event set.

$$\frac{a \in A}{(x : A \to P_x) \xrightarrow{a} P_a}$$

External Choice *External Choice* is only resolved in favor of the first process (or the second process, respectively) when the corresponding subprocess communicates a visible event. Internal steps do not influence the choice.

2.2 Timed Communicating Sequential Processes

$$\frac{P \xrightarrow{\mu} P'}{P \Box Q \xrightarrow{\mu} P'} \mu \neq \tau \qquad \frac{P \xrightarrow{\tau} P'}{P \Box Q \xrightarrow{\tau} P' \Box Q}$$

$$\frac{Q \xrightarrow{\mu} Q'}{P \Box Q \xrightarrow{\mu} Q'} \mu \neq \tau \qquad \frac{Q \xrightarrow{\tau} Q'}{P \Box Q \xrightarrow{\tau} P \Box Q'}$$

Interrupt Only terminating steps of the first process resolve the *Interrupt* in favor of it. The *Interrupt* construction is resolved in favor of the second process if it performs a visible event. Internal steps do not influence the *Interrupt*.

$$\frac{P \xrightarrow{\mu} P'}{P \triangle Q \xrightarrow{\mu} P' \triangle Q} \mu \neq \checkmark \qquad \frac{P \xrightarrow{\checkmark} P'}{P \triangle Q \xrightarrow{\checkmark} P'}$$

$$\frac{Q \xrightarrow{\mu} Q'}{P \triangle Q \xrightarrow{\mu} Q'} \mu \neq \tau \qquad \frac{Q \xrightarrow{\tau} Q'}{P \triangle Q \xrightarrow{\tau} P \triangle Q'}$$

Internal Choice *Internal Choice* is nondeterministically resolved by communicating a τ event. The environment has no influence on which process is chosen.

$$\frac{}{P \sqcap Q \xrightarrow{\tau} P} \quad \frac{}{P \sqcap Q \xrightarrow{\tau} Q}$$

Sequential Composition *Sequential Composition* is defined using two rules. The first rule is for evolving the first process of the composition with respect to non-terminating steps. The second rule is for the case where the first process terminates and the whole process evolves silently to the second process Q of the composition.

Background

$$\frac{P \xrightarrow{\mu} P'}{P;Q \xrightarrow{\mu} P';Q} \mu \neq \sqrt{} \qquad \frac{P \xrightarrow{\sqrt{}} P'}{P;Q \xrightarrow{\tau} Q}$$

Parallel Composition *Parallel Composition* ensures that events in the synchronization set A and the event $\sqrt{}$ (termination) are communicated jointly by the processes of the *Parallel Composition*. Each process can behave independently from the other in all other cases[3].

$$\frac{P \xrightarrow{\mu} P' \quad Q \xrightarrow{\mu} Q'}{P \parallel_A Q \xrightarrow{\mu} P' \parallel_A Q'} \mu \in A^{\sqrt{}}$$

$$\frac{P \xrightarrow{\mu} P'}{P \parallel_A Q \xrightarrow{\mu} P' \parallel_A Q} \mu \notin A^{\sqrt{}} \qquad \frac{Q \xrightarrow{\mu} Q'}{P \parallel_A Q \xrightarrow{\mu} P \parallel_A Q'} \mu \notin A^{\sqrt{}}$$

Hiding The *Hiding* process allows for abstracting away certain events in the behavior of a process such that these events are mapped to the single τ event.

$$\frac{P \xrightarrow{\mu} P'}{P \setminus A \xrightarrow{\mu} P' \setminus A} \mu \notin A \qquad \frac{P \xrightarrow{a} P'}{P \setminus A \xrightarrow{\tau} P' \setminus A} a \in A$$

Process Variables/ Recursion The operational rule for process variables silently unfolds its definition given by the process variable assignment *asg*.

$$\frac{X \in \mathcal{V}}{X \xrightarrow{\tau} asg(X)} \quad \text{with} \quad asg :: \mathcal{V} \Rightarrow CSP$$

---
[3] The notation $A^{\sqrt{}}$ is an abbreviation for $A \cup \{\sqrt{}\}$.

2.2 Timed Communicating Sequential Processes

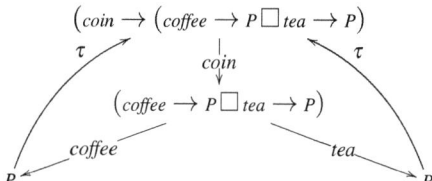

Figure 2.1: Labeled Transition System of a Coffee Machine

The presented operational semantics above facilitates the interpretation of a CSP process as an LTS. As an example, a process describing a simple coffee machine $P = coin \rightarrow (coffee \rightarrow P \square tea \rightarrow P)$ can be visualized as depicted in Figure 2.1. Note that this (recursive) process is formally defined using a process variable P and a process variable assignment asg with $asg(P) = coin \rightarrow (coffee \rightarrow P \square tea \rightarrow P)$. This process models a simple machine, which initially accepts a coin and offers the user the possibility to choose between coffee and tea. When the choice was made, the machine returns to its initial state.

Denotational Semantics and Refinement

There are several denotational semantical models of CSP. The most common denotational semantics are the trace semantics, the stable failures semantics and the failures-divergences semantics [Ros97, Sch99]. In the traces semantics of CSP, a process is represented by the set of all its possible communication sequences that only consist of visible events out of $\Sigma^{\checkmark}$. By contrast, the stable failures semantics also records the refusals of a process, i.e., the events a process can (stably) refuse after a particular communication sequence. Both these semantics are unable to recognize processes with infinite internal behavior (infinitely many consecutive τ-steps are possible), which represents defective behavior. The failures-

Background

divergences semantics fills exactly this gap by introducing the concept of divergences.

The denotational semantics of CSP is defined in terms of compositional semantical functions $\mathcal{T}$, $\mathcal{SF}$ and $\mathcal{F}/\mathcal{D}$ mapping process terms to their semantical representation. The traces of a process are given by $\mathcal{T}$, the stable failures are given by $\mathcal{SF}$, and the (possibly unstable) failures and the divergences of a process are given by $\mathcal{F}$ and $\mathcal{D}$. These semantical functions can be characterized using the operational semantics as described below. We do not present their direct definitions here.

Based on the denotational semantics, conformance in CSP can be expressed by *refinement*. Informally, $P \sqsubseteq Q$ means that Q conforms to P, or that the behaviors of Q are contained in those of P. Formally, refinement for the three denotational semantical models is defined as follows.

$$P \sqsubseteq_T Q \quad \text{iff} \quad \mathcal{T}[Q] \subseteq \mathcal{T}[P]$$
$$P \sqsubseteq_{SF} Q \quad \text{iff} \quad \mathcal{T}[Q] \subseteq \mathcal{T}[P] \wedge \mathcal{SF}[Q] \subseteq \mathcal{SF}[P]$$
$$P \sqsubseteq_{FD} Q \quad \text{iff} \quad \mathcal{F}[Q] \subseteq \mathcal{F}[P] \wedge \mathcal{D}[Q] \subseteq \mathcal{D}[P]$$

The result is that CSP provides a refinement calculus, which supports the process development stepwise from abstract specification to implementation.

Traces Semantics Let $P \stackrel{tr}{\Longrightarrow} Q$ denote that the Process P can evolve to Q using the communication sequence $tr = \langle a_1, a_2, \ldots, a_n \rangle$ with $a_i \neq \tau$, i.e.,

$$P \stackrel{\tau}{\rightarrow}^* \bullet \stackrel{a_1}{\rightarrow} \bullet \stackrel{\tau}{\rightarrow}^* \bullet \stackrel{a_2}{\rightarrow} \bullet \stackrel{\tau}{\rightarrow}^* \ldots \stackrel{\tau}{\rightarrow}^* \stackrel{a_n}{\rightarrow} \bullet \stackrel{\tau}{\rightarrow}^* Q$$

2.2 Timed Communicating Sequential Processes

Then, the traces of a process can be characterized by

$$\mathcal{T}(Proc) = \{tr \ . \ \exists Q. \ Proc \stackrel{tr}{\Longrightarrow} Q\}.$$

Note that a process is represented by a possibly infinite set of traces in the traces semantics. For example, the process $P = a \rightarrow P$ is represented by $\{\langle\rangle, \langle a\rangle, \langle a,a\rangle, \ldots\}$. In particular, the traces of a process are prefix closed, i.e., if $tr \in \mathcal{T}(P)$ and tr' is a prefix of tr (written as $tr' \leq tr$) then $tr' \in \mathcal{T}(P)$.

Stable Failures Semantics Let $P \downarrow$ abbreviate $\neg(P \stackrel{\tau}{\longrightarrow})$. This means that no internal steps are enabled in P. Additionally, let ref (refuse) denote the predicate that a process cannot engage in certain (visible) events, i.e., $P \ ref \ X$ iff $\forall a \in X \ . \ \neg(P \stackrel{a}{\longrightarrow})$. Then, the stable failures of a process can be characterized by

$$\mathcal{SF}(Proc) = \{(tr, X) \ . \ \exists Q \ . \ Proc \stackrel{tr}{\Longrightarrow} Q \wedge Q \downarrow \wedge Q \ ref \ X\}.$$

This means that the stable failures of a process contain all traces that eventually stabilize together with refused events at the end of each such trace.

Failures-Divergences Semantics For the operational characterization of the failures-divergences semantics, we first need to define a predicate that indicates that a process can possibly engage in infinitely many internal steps. A process P is said to be divergent ($P \uparrow$) if the following holds

$$P \uparrow \stackrel{def}{=} \exists \langle P_i \rangle_{i \in \mathbb{N}} \ . \ (P = P_0 \wedge \forall i \in \mathbb{N} \ . \ P_i \stackrel{\tau}{\longrightarrow} P_{i+1})$$

Then, we can define the set of divergences $\mathcal{D}(P)$ inductively.

1. if $P \stackrel{tr}{\Longrightarrow} Q \wedge Q \uparrow$ then $tr \in \mathcal{D}(P)$

Background

2. **if** $tr \in \mathcal{D}(P) \land tr \leq tr'$ **then** $tr' \in \mathcal{D}(P)$

Using this definition, the failures can be characterized by

$$\mathcal{F}(Proc) = \mathcal{SF}(Proc) \cup \{(tr,X) \, . \, tr \in \mathcal{D}(Proc)\}.$$

This means that after a divergent trace, a process can engage in every possible extension of the trace because divergent traces can be extended arbitrarily (Definition of $\mathcal{D}$). Additionally, every event can be refused after a divergent trace (since X is unbounded). This implies that divergence leads to chaotic behavior because everything can be accepted but at the same time be refused. Note that the failures-divergences semantics reduces to the stable failures semantics in the case that there are no divergent paths present.

Until now, we have presented the syntax and semantics of untimed CSP. In the following subsection, we introduce the timed extensions of Timed CSP and present its timed operational semantics. This facilitates the interpretation of Timed CSP as timed LTSs.

2.2.2 Timed CSP

Timed CSP [Sch99], in contrast to CSP, also captures the timing behavior of processes. It is possible to describe explicit delays between the occurrence of events. To this end, two new (timed) operators are included in the syntax of Timed CSP. The operational semantics of CSP is extended by rules for the timing behavior of processes. Semantical rules concerning the communication of instantaneous events are not changed.

2.2 Timed Communicating Sequential Processes

Syntax

Timed CSP shares most of the operators with (untimed) CSP. Additionally, it introduces the timed primitives $P \stackrel{d}{\rhd} Q$ (*Timeout*) and $P \triangle_d Q$ (*Timed Interrupt*) for positive real values d. Intuitively, the meaning of *Timeout* is that the process P can be triggered by some (external) event within d time units. If this happens, the *Timeout* is resolved in favor of P. If the time expires without P being triggered, process Q handles this situation, i.e., the *Timeout* is resolved in favor of Q. The *Timed Interrupt* construction has a similar meaning. Here, P can (successfully) terminate within d time units, otherwise P is aborted and Q is started. Further timed operators can be defined using these basic operators. For example the *WAIT* process is defined as $WAIT(d) \stackrel{def}{=} STOP \stackrel{d}{\rhd} SKIP$. Thus, for example in the process $WAIT(5); P$, the transition to P happens exactly after 5 time units, as $STOP$ cannot communicate any (visible) event. Based on the *WAIT* process, a delayed prefix operator can be defined as $a \rightarrow^d P \stackrel{def}{=} a \rightarrow WAIT(d); P$. Its meaning is that after communicating event a, P is reached after (exactly) d time units.

Operational Semantics

There are two main types of semantics that are typically used in the context of Timed CSP: The denotational timed failures semantics and the operational semantics, which interprets Timed CSP as timed LTS. We focus on the operational semantics in this thesis. For a presentation of the timed failures semantics, we refer to [Dav93] and [Sch99] where also complete proof systems based on the denotational semantics are presented.

The operational transitions of Timed CSP processes consist of instantaneous event transitions ($\longrightarrow$) and timed transitions ($\rightsquigarrow$). By this, Timed CSP defines a timed LTS $(TCSP, (\longrightarrow \cup \rightsquigarrow), \Sigma \cup \{\tau, \sqrt{}\}, \mathbb{R}_{>0}))$, where

Background

$\longrightarrow$ defines discrete event transitions and $\leadsto$ defines timed transitions. In the following, we introduce the operational semantics of Timed CSP. The event transitions $\longrightarrow$ for process operators already present in CSP are defined in the same way as in Section 2.2.1 and are therefore omitted here.

Basic Processes The processes *STOP*, *SKIP*, *Prefix*, and *Menu Choice* can let time arbitrarily advance.

$$\frac{}{STOP \overset{t}{\leadsto} STOP} \; t>0 \qquad \frac{}{SKIP \overset{t}{\leadsto} SKIP} \; t>0$$

$$\frac{}{a \rightarrow P \overset{t}{\leadsto} a \rightarrow P} \; t>0 \qquad \frac{}{x:A \rightarrow P_x \overset{t}{\leadsto} x:A \rightarrow P_x} \; t>0$$

Sequential Composition In the *Sequential Composition*, time can only advance if the first process cannot currently successfully terminate. This means that the corresponding event rule can be applied. In other words, termination is urgent within a *Sequential Composition*.

$$\frac{P \overset{t}{\leadsto} P' \quad \neg(P \overset{\sqrt{}}{\longrightarrow})}{P;Q \overset{t}{\leadsto} P';Q}$$

External Choice The *External Choice* process can let time advance if both of its components can.

$$\frac{P \overset{t}{\leadsto} P' \quad Q \overset{t}{\leadsto} Q'}{P \square Q \overset{t}{\leadsto} P' \square Q'}$$

Interrupt Similarly, the *Interrupt* process can let time advance if bot of its components can.

2.2 Timed Communicating Sequential Processes

$$\frac{P \stackrel{t}{\leadsto} P' \quad Q \stackrel{t}{\leadsto} Q'}{P \triangle Q \stackrel{t}{\leadsto} P' \triangle Q'}$$

Internal Choice There is no timed rule for *Internal Choice*. This means that *Internal Choice*s must be resolved immediately (by the corresponding event rule) as they occur.

Parallel Composition Just as in the case of *External Choice* and *Interrupt*, time can advance in a *Parallel Composition* when time can advance in both of its components.

$$\frac{P \stackrel{t}{\leadsto} P' \quad Q \stackrel{t}{\leadsto} Q'}{P \parallel_A Q \stackrel{t}{\leadsto} P' \parallel_A Q'}$$

Hiding The *Hiding* operator can let time only advance if the process cannot communicate an event to be hidden. In this case, the corresponding event rule for the *Hiding* operator is applicable, turning the event step into an internal τ step. This means that hidden events are performed as soon as they are possible.

$$\frac{P \stackrel{t}{\leadsto} P' \quad \forall a \in A.\ \neg(P \stackrel{a}{\longrightarrow})}{P \setminus A \stackrel{t}{\leadsto} P' \setminus A}$$

Timeout A *Timeout* can be resolved in favor of the first process when the first process communicates a visible event within time d. If the time d elapses, the *Timeout* is resolved in favor of the second process by an internal step.

Background

$$\frac{P \xrightarrow{\mu} P'}{P \stackrel{d}{\triangleright} Q \xrightarrow{\mu} P'} \mu \neq \tau \qquad \frac{P \xrightarrow{\tau} P'}{P \stackrel{d}{\triangleright} Q \xrightarrow{\tau} P' \stackrel{d}{\triangleright} Q} \qquad \frac{}{P \stackrel{0}{\triangleright} Q \xrightarrow{\tau} Q}$$

The rule for timed steps models that the "counter" d is continuously decreased.

$$\frac{P \stackrel{t}{\leadsto} P'}{P \stackrel{d}{\triangleright} Q \stackrel{t}{\leadsto} P' \stackrel{d-t}{\triangleright} Q} \, t \leq d$$

Timed Interrupt The semantics of *Timed Interrupt* is quite similar to that of *Timeout* except that the choice can only be resolved by the first process when it terminates successfully. If the time elapses, the *Timed Interrupt* is resolved in favor of the second process by an internal event.

$$\frac{P \xrightarrow{\checkmark} P'}{P \triangle_d Q \xrightarrow{\checkmark} P'} \qquad \frac{P \xrightarrow{\mu} P'}{P \triangle_d Q \xrightarrow{\mu} P' \triangle_d Q} \mu \neq \checkmark \qquad \frac{}{P \triangle_0 Q \xrightarrow{\tau} Q}$$

Again, the rule for timed steps models that the counter d is decreased.

$$\frac{P \stackrel{t}{\leadsto} P'}{P \triangle_d Q \stackrel{t}{\leadsto} P' \triangle_{d-t} Q} \, t \leq d$$

Like in CSP, Timed CSP processes can be visualized using their operational semantics. Consider the slightly extended coffee machine example from Section 2.2.1: $P = coin \rightarrow WAIT(3); (coffee \rightarrow P \square tea \rightarrow P)$. The intuition is that after a coin was inserted, the coffee machine needs 3 time units to check the validity of the coin and then it offers the choices of tea and coffee as before. The labeled transition graph is given in Figure 2.2.

2.2 Timed Communicating Sequential Processes

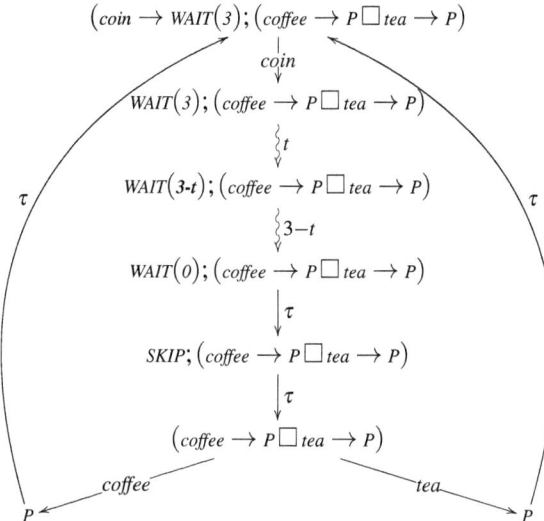

Figure 2.2: Timed Labeled Transition System of a Coffee Machine

We have omitted the time loops in the initial state and the state, where the choice is offered. Note that t can get every possible value between 3 and 0 during execution and that arbitrarily many intermediate states in between are possible. That is, the LTS becomes infinite, both in width and depth, by introducing continuously timed steps.

The event transitions of Timed CSP are best understood as communication with some environment. In particular, this means that an event is only communicated if the environment requests it. This interpretation is enforced by the semantical rules of timed steps because there is no process construction forcing a visible event (all except τ) to happen. They may be either arbitrarily delayed or their offer may be aborted by an internal step. However, internal steps are enforced to occur as soon as possible, i.e., time can advance if and only if no internal transition is enabled.

Background

2.2.3 The FDR2 Refinement Checker

All three kinds of refinement of CSP as introduced in Section 2.2.1 are supported by the automatic refinement checker FDR2 [GRA05], which proves or refutes assertions of the form $P \sqsubseteq_X Q$, for $X \in \{T, SF, FD\}$. Input processes for FDR2 are expressed in CSP_M, a machine-readable version of CSP. CSP_M expresses CSP by a small but powerful functional language, offering constructs such as *lambda* and *let* expressions and supporting pattern matching and currying. It also provides a number of predefined data types, including booleans, integers, sequences, and sets. Furthermore, user-defined data types can be defined.

A special feature of the FDR2 refinement checker is its (although limited) support for tock CSP. Tock CSP is a discretely-timed dialect of CSP. It introduces a new event *tock*, which is used to model the passage of one (discrete) time unit. The *WAIT* operator, for example, can be defined as $WAIT(d) \stackrel{def}{=}$ **if** $d = 0$ **then** $SKIP_t$ **else** $tock \rightarrow WAIT(d-1)$ under the assumption that $d \in \mathbb{N}$. The process $SKIP_t$ allows the process to terminate but also to let time advance, i.e., $SKIP_t \stackrel{def}{=} SKIP \square tock \rightarrow SKIP_t$. As shown in [Oua01], Timed CSP can be translated to tock CSP such that the verification results can be transferred back to the Timed CSP level under certain conditions. However, the semantical model of tock CSP is another than that of standard CSP. This means, especially, that τ needs to be preferred over *tock* in order to gain τ-urgency. Support for this special semantics is given by the τ-priority model of FDR2 such that FDR2 can be used to verify translated Timed CSP processes. Currently the τ-priority model is only supported for the traces semantics. Due to this fact, we may currently only consider traces equivalence in FDR2 in the context of our framework. In future releases of FDR2, other semantical model will be supported as well.

In this section, we have given a brief introduction to the operational semantics of CSP and Timed CSP. The operational semantics facilitates the interpretation of (Timed) CSP as a (timed) LTS. This is important because it enables us to instantiate the notions of bisimulation in the context of Timed CSP. Based on the operational semantics of CSP, it is possible to characterize the denotational semantics of CSP. These are used in the FDR2 refinement checker, which allows for checking refinement between (untimed) processes. Furthermore, it is also possible to verify Timed CSP processes with FDR2 to a certain degree. To this end, Timed CSP needs to be discretized to a dialect of CSP, which is called tock CSP. We have adapted and extended an existing discretization approach, which is explained in Chapter 6. Besides FDR2, we also employ the UPPAAL model checker within our framework. The background of timed automata is summarized in the next section.

2.3 UPPAAL Timed Automata

The formalism of timed automata was introduced in [AD94]. It is an extension of finite automata that adds clock variables. Clocks can be manipulated and be used in clock constraints such that precise timed behavior can be modeled. Concurrent systems are modeled using networks of timed automata. This means that each process is modeled as a single timed automaton. The single automata are then composed in a network. The processes of a network are executed in an interleaving semantics and can communicate by binary handshake synchronization on events or by broadcast communication. As an important fact, the continuous timed operational semantics of timed automata can be discretized using region graphs or zone graphs. This makes it possible to employ automatic verification tools such as the UPPAAL model checker [BY04] to perform verification tasks fully automatically.

Background

2.3.1 Syntax

A timed automaton consists of a finite set of locations, a finite set of directed edges between locations and a finite set of clock variables. Each location is equipped with a set of clock constraints called location invariants. The intuition for invariants is that the automaton may stay in a particular location as long as the invariants are satisfied by the current clock valuation. During execution, locations can be changed by taking discrete transitions. A transition corresponds to an edge of the automaton. Each edge is equipped with an event, a finite set of transition guards and a finite set of clocks that are reset to 0 when taking the transition.

Clock constraints $B(C)$ over a set C of clock variables are finite conjunctions of atomic clock constraints. These are of the form $x \sim n$ or $x - y \sim n$ where $x, y \in C$, $\sim \in \{\leq, <, =, \geq, >\}$, and $n \in \mathbb{N}$.

Definition 12 (Timed Automaton) *Formally, the syntax of a timed automaton is given by a tuple $(L, l_0, C, \Sigma, E, I)$ where*

- *L is a finite set of locations and $l_0 \in L$ is the initial location.*
- *C is a finite set of clock variables.*
- *Σ is a finite set of actions with $\tau \in \Sigma$ denoting an internal transition.*
- *$E \subseteq L \times B(C) \times \Sigma \times \mathbb{P}(C) \times L$ is a finite set of edges. We also use the notation $l \xrightarrow{(g,a,r)} l'$ for $(l, g, a, r, l') \in E$.*
- *$I :: L \Rightarrow B(C)$ is a function assigning an invariant to each location.*

As a an example of a simple timed automaton consider Figure 2.3. It models a simple producer process, which can produce something, send the product to someone else, and clean the working place. The automaton consists of three locations (*initial, produce, clean*). The initial location is marked with ⓞ. The automaton may stay for at most three time units in

2.3 UPPAAL Timed Automata

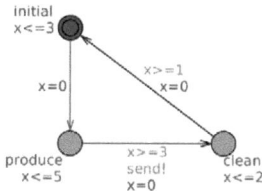

Figure 2.3: A Simple Timed Automaton

its initial location. Then, it produces something in location *produce*, which takes between three and five time units. The location *produce* is left by the transition that sends the product to someone else. The reached location is *clean*, where between one and two time units pass before the automaton reaches its initial location again.

2.3.2 Semantics

The semantics of a single timed automaton $(L, l_0, C, \Sigma, E, I)$ and networks of timed automata is given in terms of a continuously timed operational semantics. We start with the operational semantics of a single timed automaton, which defines a timed LTS $(S, \longrightarrow, \Sigma, \mathbb{R}_{\geq 0})$. The set of states $S = L \times \mathbb{R}_{\geq 0}^{C}$ of a timed automaton consists of tuples (l, u) where $l \in L$ is a location and $u :: C \Rightarrow \mathbb{R}_{\geq 0}$ is a clock valuation. For all clocks in a timed automaton time advances at the same speed. Thus, the clock valuations u change in the form $u + d$ where all clocks are advanced with the same difference $d \in \mathbb{R}_{\geq 0}$. This means that $u + d = \lambda c. u(c) + d$. There are two sorts of steps in the operational semantics: (1) instantaneous transition steps and (2) timed steps.

Definition 13 (Semantics of a (single) Timed Automaton)

(1) $(l, u) \xrightarrow{a} (l', u')$ iff $\exists g\ r.\ l \xrightarrow{(g,a,r)} l' \wedge u' = [r \mapsto 0]u \wedge u \models g$
$\wedge u' \models I(l')$

Background

(2) $(l,u) \xrightarrow{d} (l,u+d)$ iff $\forall d'.\, 0 \leq d' \wedge d' \leq d \longrightarrow u+d' \models I(l)$

The abbreviation $[r \mapsto 0]u$ stands for $\lambda c.$ **if** $c \in r$ **then** 0 **else** $u(c)$. Thus, all clocks in r are reset to 0. To express that a clock valuation u satisfies a clock constraint g, the notation $u \models g$ is used. This expression is evaluated by replacing the clock variables in g according to their values in u.

A network of timed automata is the parallel composition of n timed automata $\mathcal{A}_i = (L_i, l_0^i, C, \Sigma, E_i, I_i)$. This means that they share the same clocks and events. Again the operational semantics of networks of timed automata defines a timed LTS $(S, \longrightarrow, \{\tau\}, \mathbb{R}_{\geq 0})$. The set of states is given by $S = (L_1 \times L_2 \times \cdots \times L_n) \times \mathbb{R}^C_{\geq 0}$. This means that a state $(\bar{l}, u)$ consists of a location vector $\bar{l}$ of length n and a clock valuation u. The transition relation is defined below. We assume that each label $l \in \Sigma$ is of the form $c?$, $c!$ or τ. Then, the operational semantics of a network of timed automata is given in terms of three possible steps: (1) instantaneous independent transition steps, (2) instantaneous transition steps modeling synchronization between two automata and (3) timed steps.

Definition 14 (Semantics of a Network of Timed Automata)

(1) $(\bar{l}, u) \xrightarrow{\tau} (\bar{l}[l_i'/l_i], u')$
    iff $\exists g\, r.\, l_i \xrightarrow{\tau, g, r} l_i' \wedge u' = [r \mapsto 0]u \wedge u \models g \wedge u' \models I(\bar{l}[l_i'/l_i])$

(2) $(\bar{l}, u) \xrightarrow{\tau} (\bar{l}[l_i'/l_i, l_j'/l_j], u')$
    iff $\exists g_i\, r_i\, g_j\, r_j\, c.\, i \neq j \wedge l_i \xrightarrow{c!, g_i, r_i} l_i' \wedge l_j \xrightarrow{c?, g_j, r_j} l_j'$
    $\wedge\, u' = [r_i \cup r_j \mapsto 0]u \wedge u \models g_i \wedge g_j \wedge u' \models I(\bar{l}[l_i'/l_i, l_j'/l_j])$

(3) $(\bar{l}, u) \xrightarrow{d} (\bar{l}, u+d)$
    iff $\forall d'.\, 0 \leq d' \wedge d' \leq d \longrightarrow u+d' \models I(\bar{l})$

2.3 UPPAAL Timed Automata

In this definition, the global invariant I is defined to be the conjunction of all local invariants, i.e., $I(\langle l_1,\ldots,l_n\rangle) = I_1(l_1) \wedge \cdots \wedge I_n(l_n)$. Furthermore, $\bar{l}[l'_i/l_i]$ denotes the vector $\bar{l}$ with l_i replaced by l'_i at position i.

The model of timed automata described so far is extended in UPPAAL by urgent and committed locations, datatypes and data variables that can, for example, be used in transition guards, broadcast communication and urgent channels. In urgent locations no time may advance, this means that they must be left at the same point in time as they are entered. For committed locations, the same holds. Additionally, in a network of timed automata, committed locations get priority over other locations. This means that in a network of timed automata, all processes have to leave their committed locations before they can leave other locations. For our purposes, urgent locations and data variables are of particular interest.

In Figure 2.4, a network of three timed automata is shown. It consists of the producer process discussed above, a consumer and a buffer. When the producer has finished the product, it sends it to the buffer, which changes its state and updates the global variable *empty*. The consumer stays in its initial location between one and two time units. Then it goes to the urgent location *lookup* marked with ◎. If the buffer is empty, the consumer goes back into its initial location. Otherwise, the edge labeled with *receive* is enabled such that after performing this edge, the consumer is in location *sell* and the buffer goes back into its initial location. The consumer is selling the product for one to three time units before it also goes back into its initial location.

In the following, we briefly describe the capabilities of the UPPAAL tool suite, focusing especially on the model checking capabilities.

Background

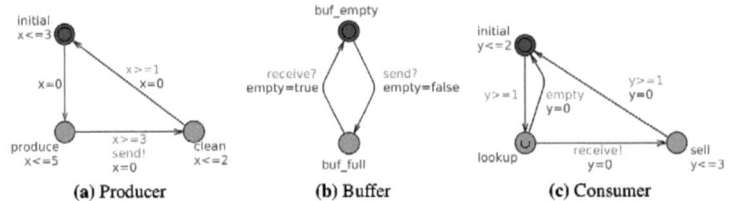

Figure 2.4: A Producer-Consumer System

2.3.3 The UPPAAL Tool Suite

UPPAAL [BY04] is a tool suite for modeling, simulation and verification of (networks of) timed automata. Properties of timed automata are expressed in a restricted version of CTL. Based on the zone graph construction, these formulae are verified using model checking techniques. In the case that a formula is not satisfied by a given timed automata model, the user is provided with a counterexample in terms of a path through the system leading to a state showing the invalidity. This path can be visualized using the UPPAAL simulator.

Formulae in UPPAAL's query language are built upon state formulae and path formulae. State formulae can be evaluated in a particular state. For example, it can be queried whether a certain variable x is below 4 by $x \leq 4$, that an automaton of the network is in a certain location $A1.init$, or that the state is deadlocked, i.e., there are no outgoing (instantaneous) transitions. These basic state formulae can be combined using the usual logical connectives such as $\wedge$ and $\neg$.

Path formulae are used to express properties about possible executions in the timed automaton. There are four constructors for path formulae. Let ϕ and ψ be state formulae then $E\Diamond\phi$, $E\Box\phi$, $A\Diamond\phi$, $A\Box\phi$, and $\phi \rightsquigarrow \psi$ are path formulae. In contrast to CTL, path formulae may not be nested in

2.3 UPPAAL Timed Automata

the UPPAAL query language. The meaning of the path formulae is the following.

- $E\Diamond\ \phi$ – there exists a path such that eventually ϕ holds
- $E\Box\ \phi$ – there exists a path such that always ϕ holds
- $A\Diamond\ \phi$ – for all paths eventually ϕ holds
- $A\Box\ \phi$ – for all paths ϕ always holds
- $\phi \leadsto \psi$ – whenever ϕ then eventually ψ holds

The last formula corresponds to the CTL formula $A\Box\ (\phi \longrightarrow A\Diamond\ \psi)$.

Formulae are interpreted with respect to the initial state of a network of timed automata. It is given by $(\langle l_0^1, l_0^2, ..., l_0^n \rangle, u_0)$ for the network of n timed automata $\mathcal{A}_i = (L_i, l_0^i, C, \Sigma, E_i, I_i)$ where u_0 maps each clock to 0. To express, for example, that a timed automaton cannot deadlock the formula $A\Box\ \neg deadlock$ can be used. To express that each received *a* event is eventually answered by a *b* event within 3 time units, the formula $P.a\_received \leadsto P.b\_received \land x \leq 3$ can be used. This means that the information that a special event happened has to be encoded into the locations of the timed automaton. Furthermore, it is assumed that each entering transition to state *P.a_received* sets the clock variable x to 0.

Model checking is realized on the zone graph of a network of timed automata. This is a finite abstraction of timed automata, which is based on the assumption that only finite integers are used in clock constraints. Under this assumption, two clock valuations u and v can be considered equivalent if the following three properties hold for all clocks x and y.

- the integer parts of $u(x)$ and $v(x)$ are identical
- the fractional part of $u(x)$ is 0 *iff* the fractional part of $v(x)$ is 0

Background

- the fractional part of $u(x)$ is less or equal to the fractional part of $u(y)$ *iff* this holds for the fractional parts with respect to v

Furthermore, there is a global maximal constant up to which these properties have to hold. This constant is given by the maximal constant occurring in the timed automaton and the formula under consideration. The equivalence sketched above induces region-equivalence of clock valuations. This equivalence can be optimized in order to gain zone-equivalence. For a thorough presentation, we refer to [BY04]. The important fact is that there are only finitely many regions and zones to be considered for a given network of timed automata. Therefore, a finite abstract LTS can be constructed on which model checking can be applied.

The example given in Figure 2.4 fulfills, for example, the following properties.

- $A\square \neg deadlock$
- $true \leadsto Producer.produce$
- $Producer.produce \leadsto Consumer.sell$
- $A\square(Producer.produce \wedge Producer.x = 5 \longrightarrow empty)$

The first property states that there is no state without possible outgoing discrete transitions. The second states that location *Producer.produce* is always eventually reached. The third property expresses that a produced product is eventually sold. Finally, the last property states that the buffer is ensured to be empty after at most five time units when the product is produced. Note that all except the second of these properties would not hold anymore if the selling of a product in the *Consumer* automaton was allowed to take, for example, at most six time units. Then, a situation could arise in which the producer wants to send the product to the buffer but cannot because the buffer was not emptied in time and the consumer has currently no possibility to do so. This situation would cause the system to deadlock.

In this section, we have introduced the dialect of UPPAAL timed automata and presented the capabilities of the UPPAAL tool suite. UPPAAL together with the FDR2 refinement checker (see Section 2.2.3) build the basis for the verification of instances of parameterized real-time systems within our framework. The comprehensive verification is performed in the Isabelle/HOL theorem prover. In the next section, the basic features of this theorem prover are presented.

2.4 The Isabelle Theorem Prover

Isabelle is a generic interactive proof assistant. It enables the formalization of mathematical models and the mechanical verification of theorems about them. There are two main advantages of using a theorem prover like Isabelle. First, mathematical proofs are machine-checked, which implies that corner cases cannot be overlooked. Second, proofs can be partly automatized by using tactics. Isabelle can be instantiated with different object logics. One instantiation is Isabelle/HOL [NPW02], which is based on Higher Order Logic. The main advantage of HOL is its very high expressive power. Theorem provers based on HOL require a high level of expertise but enable reasoning about models whose state space is too large to be automatically checked by, for example, a model checker. Unlike applying automatic verification techniques, proving theorems in a theorem prover like Isabelle/HOL is highly interactive. Specifications have to be designed carefully to allow for the verification of properties.

To briefly describe higher-order logic, we shall start with second-order logic. Second-order logic is an extension of first-order predicate logic in that it not only allows quantifiers expressing "for all objects (in the uni-

Background

verse of discourse)" but also "for all properties on objects (in the universe of discourse)". This means that, for example, the induction rule for natural numbers can be expressed as $\forall P.\ P\ 0 \wedge \forall k.\ (P\ k \longrightarrow P\ k+1) \longrightarrow \forall n.\ P\ n$ where quantification over the first-order predicate P is used. Third-order logic then allows quantification over second-order predicates and so on. Higher-Order Logic generalizes this concept by allowing quantification over predicates and functions of arbitrary order.

In a nutshell, Isabelle/HOL is the combination of functional programming and logic. The user can define inductive datatypes and functions concisely. Afterwards, the user can prove properties about them using for example induction and automatic tactics, which try to solve proofs automatically. Tactics try to apply the basic rules of the logic in a sophisticated order. As HOL is generally undecidable, these tactics are not complete and thus require user interaction in guiding the proof structure.

In the following, we give an overview of Isabelle's modeling and verification capabilities that are used in this thesis. We start with explaining the general structure of Isabelle theories. Then, we explain which kinds of definitions can be introduced for describing formal models. Furthermore, we describe how proofs can be performed to show properties of them. Finally, we present the concept of locales in Isabelle and how they can be used to structure theories in a modular way.

2.4.1 Structure of Isabelle Theories

Isabelle enables the development of user-defined theories and allows existing theories to be imported. Thereby, definitions and lemmas of the imported theories can be reused in user-defined definitions and proofs. The common structure of an Isabelle theory is sketched in Figure 2.5. Here, the new theory myTheory is based on existing theories T1, T2, ..., Tn. Then, the theory is typically divided into two parts: a definition part and

2.4 The Isabelle Theorem Prover

```
theory myTheory
imports T1 T2 ... Tn
begin
(* definition part *)
  definition d1 ...
  definition d2 ...
  ...

(* verification part *)
  lemma l1: ...
  lemma l2: ...
  ...
end
```

Figure 2.5: Common Structure of an Isabelle-Theory

a verification part. The definition part is used to introduce new syntax (by means of, for example, datatypes, functions and sets). In the verification part, properties are formulated over the newly introduced definitions and verified using Isabelle's proof machinery. By giving names to lemmas, they can be reused later in the new theory or in theories importing it.

2.4.2 Performing Proofs in Isabelle

Isabelle's meta logic is an intuitionistic higher-order logic. Its basic logical operators are implication ($\Longrightarrow$), equality ($\equiv$), and universal quantification ($\bigwedge$). The general structure of a meta logical proposition that is used for representing a proof state (also called proof goal) is

$$\bigwedge x_1 \ldots x_n. \; [\![A_1 \,;\, \ldots \,;\, A_m]\!] \Longrightarrow P$$

$[\![A_1 \,;\, \ldots \,;\, A_m]\!] \Longrightarrow P$ denotes $A_1 \Longrightarrow (A_2 \Longrightarrow (\ldots \Longrightarrow (A_m \Longrightarrow P) \ldots))$. Such a proof state can be read as "for fixed variables $x_1 \ldots x_n$ and under the assumptions $A_1 \ldots A_n$, property P holds". The terms $A_1, \ldots, A_m, P$ are usually terms in an object logic of Isabelle. These consist of operators

Background

and axioms defining the logic. Axioms are formulated as proof rules as explained in the following.

The common structure of a proof rule is

$$[\![B_1 ; \ldots ; B_m]\!] \Longrightarrow Q$$

In contrast to proof goals, locally fixed variables are not present here any more. Instead, they are replaced by schematic variables (also called meta variables) $?v$. This means that when some axiom is introduced or some lemma 1 was proved successfully, all fixed variables (free variables) are replaced by schematic variables in the resulting proof rule. Unification is enabled by schematic variables. When applying a rule to a proof goal, these schematic variables are automatically unified by Isabelle if possible. If there is more than one possibility, the unification process can be guided. If a unification succeeds and the schematic variable occurs in the other terms of the applied rule, they are replaced accordingly.

If the conclusion of a proof goal occurs in its assumptions, the subgoal can be finished by Isabelle's `assumption` command. Axioms and proven lemmas follow the structure of proof goals and can be applied if unification succeeds. Let, for example, be r such an axiom or verified lemma called rule in the following with $[\![B_1; B_2; \ldots; B_m]\!] \Longrightarrow Q$.

The rule r can be applied in a backward manner by `apply(rule r)` if P and Q can be unified. Then, new subgoals arise, which require that the assumptions of the applied rule are implied by the assumptions of the current proof goal, i.e., it has to be shown that[4]

---

[4]Note that we omit the locally fixed variables in the proof state in the following.

2.4 The Isabelle Theorem Prover

$$[\![A_1\,;A_2\,;\ldots\,;A_n]\!] \implies B_1$$
$$\ldots$$
$$[\![A_1\,;A_2\,;\ldots\,;A_n]\!] \implies B_m$$

Rules can also be applied in a forward manner by using Isabelle's `drule` command. If some A_i can be unified with B_1, the resulting proof goals after the command `apply(drule r)` are

$$[\![A_1\,;\ldots A_{i-1}\,;A_{i+1}\ldots\,;A_n]\!] \implies B_2$$
$$\ldots$$
$$[\![A_1\,;\ldots A_{i-1}\,;A_{i+1}\ldots\,;A_n]\!] \implies B_m$$
$$[\![A_1\,;\ldots A_{i-1}\,;A_{i+1}\ldots\,;A_n\,;Q]\!] \implies P$$

In the last subgoal, assumption A_i is deleted due to the use of `drule`. If this assumption shall be kept, `frule` can be used instead.

A combination between forward and backward reasoning is provided by `erule`. If some A_i can be unified with B_1 and P can be unified with Q, then the following subgoals remain after the command `apply(erule r)`

$$[\![A_1\,;\ldots\,;A_{i-1}\,;A_{i+1}\,;\ldots A_n]\!] \implies B_2$$
$$\ldots$$
$$[\![A_1\,;\ldots\,;A_{i-1}\,;A_{i+1}\,;\ldots A_n]\!] \implies B_m$$

Background

For large proofs, it would be very cumbersome to apply one single rule after another. Therefore, Isabelle provides the possibility to automatize proofs by tactics. When these are applied, a set of rules is applied in a "useful" order and may depend on the structure of given terms in the current proof goal(s). Important tactics are, for example, simp, clarify, auto and blast. The tactic simp is a simplification method, which rewrites subgoals by rules of the form $t = s$, and thereby replaces occurrences of t by s. The method clarify applies safe logical rules to the subgoal without simplifying them and without splitting subgoals. Thereby, for example, existential quantifiers are eliminated in the assumptions and universal quantifiers are eliminated in the conclusion of a subgoal. The auto method combines classical reasoning with simplification. The method blast is a classical tableau reasoner. The proofs derived by this method are reconstructed in Isabelle using usual logical inferences. There are a lot more tactics implemented in Isabelle. For a thorough overview we refer to [Wen11].

2.4.3 Datatypes, Functions and (Co)inductive Sets

Isabelle/HOL provides a polymorphic type system. The basic types of Isabelle are predefined types like bool, nat or real and type variables typically written as 'a, 'b, 'c and so on. These basic types are used to define function types, i.e., if t_1 and t_2 are types, then $t_1 \Rightarrow t_2$ is also a type. Note that this especially enables the definition of higher-order function types.

Inductive Datatypes

To introduce user-defined types, datatypes can be defined similarly as in functional programming languages. For example, the predefined datatype of lists is introduced as follows.

2.4 The Isabelle Theorem Prover

```
datatype 'a list = Nil            ("[]")
               | Cons "'a" "'a list"  (infixr "#")
```

The datatype is parameterized with the type variable 'a such that lists of arbitrary (fixed) type can be defined. It provides two constructors: one for the empty list and one for building new lists from a single element and an existing list. The terms in brackets introduce a special syntax for the constructors. Thus, [] denotes the empty list Nil and # can be used to describe composed list with an infix notation. This means that for example x#y#xs denotes the list Cons x (Cons y xs).

When defining datatypes in Isabelle, common lemmas are proved automatically that provide, for example, case-distinction and induction rules. The proof rule for induction over lists is

list.induct: $[\![\ ?P\ []\ ;\ \bigwedge x\ xs.\ ?P\ xs \Longrightarrow ?P(x\#xs)\ ]\!]$
              $\Longrightarrow ?P\ ?1$

To prove that a certain property ?P holds for arbitrary lists ?1, it has to be shown that the predicate holds for the empty list and that under the assumption that the predicate holds for some list xs, it also holds for a list *x#xs*. Remember that variables preceded with ? are called schematic variables. These are free variables, which can be instantiated arbitrarily when applying the rule to a proof goal.

Syntactical Abbreviations

Isabelle allows for the introduction of syntactical abbreviations to make definitions more readable. For example, such a syntactical abbreviation is introduced above for the list constructor Cons. Using the mechanism of syntactical abbreviations more directly, we also could have introduced the following.

Background

```
abbreviation Cons_syntax :: 'a ⇒ 'a list ⇒ 'a list (infixr "#")
where "x#xs ≡ Cons x xs"
```

Abbreviations can be used to manipulate Isabelle's parser to some degree. Whenever the term `Cons x xs` occurs in some Isabelle script, the Isabelle parser rewrites it to `x#xs` but internally, the term `Cons x xs` is worked on. Note that this mechanism also makes it possible to introduce special syntax for already existing definitions. This makes the syntax within Isabelle proofs very readable such that the proof goals can be interpreted in a very intuitive way.

Functions

The simplest way to introduce a new function is to define it non-recursively. For example, a predicate for empty lists can be defined as follows.

```
definition empty :: "'a list ⇒ bool"
where empty l ≡ (l = [])
```

Due to the polymorphic type of `empty`, the predicate can be applied to lists of arbitrary type. This means that `empty` can be applied on lists of reals as well as on lists of functions from nat to nat and so on.

As common in functional programming, Isabelle also allows for the definition of recursive functions. However, it must be ensured that a defined function is always terminating. Otherwise, inconsistencies would be introduced into the logics of Isabelle. To define, for example, a function calculating the length of a list, we can use the `primrec` command defining a primitive recursive function. For these kinds of functions, Isabelle automatically provides a well-founded ordering on the arguments, which is used for proving termination.

2.4 The Isabelle Theorem Prover

```
primrec length :: "'a list ⇒ nat"
where
  "length [] = 0"
| "length (x#xs) = length xs + 1"
```

Pattern matching is allowed for the constructors of a type defined with datatype. Furthermore, in the recursive call it is only permitted to apply the function to be defined on direct child nodes of the constructor pattern.

In the case that these restrictions are too strong to define a function, Isabelle provides the possibility to define arbitrary recursive functions. Then, termination has to be manually shown if a well-founded lexicographic ordering cannot be found automatically. To this end, Isabelle supports to define functions using the fun and function commands. If fun is used, Isabelle tries to find a lexicographic ordering automatically. In the case of function, the user has to provide a well-founded ordering and prove that for each recursive call the arguments become "smaller" with respect to that ordering.

(Co)inductive Sets

Isabelle has very well developed theories for sets and allows (co)inductive sets to be defined in a convenient way. Consider the following inductive definition of positive even integers.

```
inductive_set even :: "int set"
where
  "0 ∈ even"
| "n ∈ even ⟹ n+2 ∈ even"
```

As in the case of datatypes Isabelle automatically performs a proof providing an induction scheme for inductive sets. For even it is

```
even.induct:   ⟦ ?x∈N1 ; ?P 0 ;
                ⋀ n.⟦ n∈N1 ; ?P n ⟧ ⟹ P(n+2) ⟧
              ⟹ ?P ?x
```

Background

Isabelle also provides the possibility to define coinductive sets. This means that instead of a least fixed point a greatest fixed point is taken for interpreting the introduction rules of the set.

Let us consider the example from above using a greatest fixpoint interpretation. The only difference of even2 with respect to even is that Isabelle's coinductive_set command is used instead of inductive_set.

```
coinductive_set even2 :: "int set"
where
  "0 ∈ even2"
| "x ∈ even2 ⟹ x+2 ∈ even2"
```

Then, the corresponding coinduction scheme is

even2.coinduct: ⟦ ?X ?x ; ⋀ x.?X x ⟹ x=0 ∨
              (∃ n.x=n+2 ∧ (?X n ∨ n ∈ even2)) ⟧
              ⟹ ?x ∈ even2

The common goal of coinduction is to show that some element is in the defined set whereas induction is used to show that a certain property holds for all the elements of an inductively defined set. To verify that some ?x is in the set even2, a predicate ?X containing ?x has to be found. Furthermore, it has to be shown that ?X does not disagree to the given rules. This means that each element x satisfying ?X is either 0 or some $n+2$ with n also satisfying ?X or being already in the set even2. Note that ?X can be thought of a set since the type 'a set is a type synonym of $'a \Rightarrow bool$. By taking ?X as the set UNIV (the set containing all elements of a particular type), we can easily show that each integer number is contained in even2, because UNIV does not disagree to the given rules of even2.

2.4 The Isabelle Theorem Prover

lemma cantor: "∀ f:: 'a ⇒ 'a set. ¬ surj f"	1. ∀ f. ¬ surj f
apply(rule allI)	1. ⋀ f. ¬ surj f
apply(rule notI)	1. ⋀ f. surj f ⟹ False
apply(unfold surj_def)	1. ⋀ f. ∀ y. ∃ x. y=f x ⟹ False
apply(erule_tac x="{x. x ∉ f x}")	1. ⋀ f. ∃ x. {x. x ∉ f x}=f x ⟹ False
apply(erule exE)	1. ⋀ f x. {x. x ∉ f x}=f x ⟹ False
apply(case_tac "x ∈ f x")	1. ⋀ f x. ⟦ {x. x ∉ f x}=f x ; x ∈ f x ⟧ ⟹ False
	2. ⋀ f x. ⟦ {x. x ∉ f x}=f x ; x ∉ f x ⟧ ⟹ False
apply(blast)	1. ⋀ f x. ⟦ {x. x ∉ f x}=f x ; x ∉ f x ⟧ ⟹ False
apply(blast)	No subgoals!
done	

Figure 2.6: Proof of Cantor's Theorem in "apply" Notation

2.4.4 Writing Down Proofs

Basically, there are two possibilities for writing down proofs. The first is to use the "apply-style" with which the current proof goal is manipulated directly. The second is the Isar-style that enables the visualization of proofs such that they resemble proofs in mathematical text books and can therefore be better understood even without Isabelle being started. The small examples in Figure 2.6 and Figure 2.7 show two proofs of Cantor's theorem stating that there is no surjective function from a set to its powerset, or in other words, that the cardinality of a powerset is strictly larger than the cardinality of its basic set. The "apply-style" proof is hard to understand without the corresponding subgoals. However, the "Isar-style" proof is more self-explanatory and also more robust when parts of the proof are changed. The price that has to be paid for this is that developing a human-readable proof takes far more time than to develop an apply-style proof.

Background

```
lemma cantor:
  shows "∀ f :: 'a ⇒ 'a set . ¬ surj f"
proof
   fix f::"'a ⇒ 'a set"
   show "¬ surj f"
   proof
     assume sg: "surj f"
     then obtain x where diag_x: "{y . y ∉ f y} = f x"
                     unfolding surj_def by(blast)
     thus False
     proof(cases "x ∈ f x")
     case True
        from diag_x have "x ∉ f x" by blast
        thus False using prems by blast
     next case False
        from diag_x have "x ∈ f x" by blast
        thus False using prems by blast
     qed
   qed
qed
```

Figure 2.7: Proof of Cantor's Theorem in Isar Notation

2.4.5 Modular Verification with Locales

We make use of Isabelle's locale mechanism [KWP99, Bal06] to structure our theories presented in Chapter 5. Locales are used to define reusable proof contexts. They are built upon a set of typed function symbols and a set of assumptions with respect to the declared function symbols. In the context of a particular locale, abstract lemmas can be shown using the corresponding function symbols and assumptions. The advantage of locales is that the locale assumptions need not be included into the assumptions of each lemma. Instead, they are available when performing proofs in the context of a locale. Later, when locales are interpreted for concrete functions and the assumptions are shown to be fulfilled by the respective interpretation, the abstract lemmas of the locale are inherited. This means, for example, that a locale for lattices may be defined and general fixpoint theorems are verified abstractly. Later, when the locale is interpreted for, say, the standard lattice on sets, the fixpoint theorems need not be shown again

2.4 The Isabelle Theorem Prover

```
locale my_locale =           context my_locale
    fixes d1::  ...          begin
      and d2::  ...            ...
                              lemma 1:  ...
                                ...
    assumes a1:  ...            apply(rule a1)
        and a2:  ...            ...
        ...                   done
                                ...
                              end
```

Figure 2.8: Structure of Locales

as they have already been proved abstractly. This means that proofs can be modularized and respective lemmas can be reused in different contexts.

The general structure of a locale is shown in Figure 2.8. There, a locale with name `my_locale` is introduced containing function symbols d1, d2, ... and assumptions a1, a2, ... In the `context` of the locale, lemmas can be proved using the function symbols and the assumptions.

As an example, we consider locales for orderings and lattices following the examples of [Bal10]. A partial order consists of a relation $\sqsubseteq$ satisfying reflexivity, antisymmetry and transitivity. This can be defined in terms of a locale `partial_order` as follows.

```
locale partial_order =
fixes le :: "'a ⇒ 'a ⇒ bool" (infixl "⊑" 50)
assumes refl: "x ⊑ x"
    and anti_sym: "⟦ x ⊑ y ; y ⊑ x ⟧ ⟹ x=y"
    and trans: "⟦ x ⊑ y ; y ⊑ z ⟧ ⟹ x ⊑ z
```

In the context of this locale, new definitions can be introduced and verified.

Background

```
context partial_order
begin
  definition less :: 'a ⇒ 'a ⇒ bool (infixl "⊏" 50)
  where
    "x ⊏ y ≡ (x ⊑ y ∧ x ≠ y)"

  lemma less_le_trans: "⟦ x ⊏ y ; y ⊑ z ⟧ ⟹ x ⊏ z"
    by(unfold less_def, blast intro: trans anti_sym)
end
```

In the proof of `less_le_trans`, we make use of the `trans` rule, which is an assumption of the locale.

Extending Locales

Existing locales can be extended by enriching them with new function symbols and additional assumptions. Lemmas that have been proved in the original locale are also available in the extended one.

The locale of partial orders can be extended to define total orders and lattices.

```
locale total_order = partial_order +
assumes total: "x ⊑ y ∨ y ⊑ x"
```

```
locale lattice = partial_order +
assumes ex_inf: "∃ i. i⊑x ∧ i⊑y ∧ (∀ z. z⊑x ∧ z⊑y ⟶ z⊑i)"
    and ex_sup: "∃ s. x⊑s ∧ y⊑s ∧ (∀ z. x⊑z ∧ y⊑z ⟶ s⊑z)"
```

Sublocales

In the preceding example, total orders and lattices are defined independently. To express that each total order is also a lattice, the `sublocale` command can be used.

2.4 The Isabelle Theorem Prover

```
sublocale total_order ⊆ lattice
  apply(unfold_locales , insert total)
  apply(blast intro: refl)+
done
```

To verify this sublocale relation, it is necessary to show that the assumptions of `total_order` imply the assumptions of `lattice`. This means that it has to be shown that there exists an infimum and a supremum for each pair of x and y. As in this case either $x \sqsubseteq y$ or $y \sqsubseteq x$ holds due to the property of total orders, the infimum is given either by x or by y. The same argumentation holds for suprema.

Locale Interpretation

A locale can be instantiated by providing concrete functions and proofs for the locale assumptions. For example, we can prove that $\leq$ defined on the natural numbers is a total order.

```
interpretation nat: total_order "op ≤::nat ⇒ nat ⇒ bool"
  by(unfold_locales , auto)
```

Thereby, all definitions and lemmas for partial orders, total orders and lattices are inherited for the natural numbers automatically.

In this section, we have introduced some of the main features of Isabelle. We explained how Isabelle theories are structured and how mechanical proofs can be performed. To introduce new definitions, the mechanisms of Isabelle datatypes, functions and (co)inductive sets can be used. One of the advantages of Isabelle is that syntactical abbreviations can be introduced to make Isabelle theories more readable. To structure abstract theories in Isabelle, locales are a very convenient possibility. They enable the definition of proof contexts that can be extended, shown to be interpretable in another proof context, or interpreted by a concrete instantiation. In all cases, proofs that have been shown in the respective proof

Background

context are inherited. Based on the expressive HOL, Isabelle/HOL facilitates the verification of properties of infinite systems using, for example, induction and coinduction. Altogether, this makes Isabelle a very powerful and convenient proof environment.

2.5 Summary

In this chapter, we have presented the relevant background of our framework for the mechanical verification of parameterized real-time systems. We have introduced labeled transition systems in general and the notion of bisimulation based on this. To express properties of labeled transition systems, we have introduced (untimed) Hennessy-Milner logic. Then, we have introduced the semantics of (Timed) CSP focusing on the operational semantics as this allows (Timed) CSP to be interpreted as a (timed) labeled transition system. Another formalism for modeling and verifying real-time systems is timed automata. We have introduced a dialect of timed automata, which is used in the UPPAAL tool suite, and have presented the capabilities of this tool suite. Finally, we have given a brief introduction to the Isabelle/HOL theorem prover.

In the next chapter, we discuss related work of this thesis. In Chapter 4, we present our framework for the mechanical verification of parameterized real-time systems. There, we especially show how the presented models and verification tools are integrated into our framework.

3 Related Work

In this chapter, we discuss related work of this thesis. In our framework, which is presented in detail in the next chapter, we consider the automatic verification of (finite) instances of parameterized systems modeled in Timed CSP and the comprehensive verification of parameterized real-time systems in a theorem proving environment. Therefore, this chapter is structured as follows: In Section 3.1, we discuss approaches for the automatized verification of real-time systems focusing especially on the automatized verification of Timed CSP. Additionally, we discuss approaches for the logical description of requirements for real-time systems. As our verification framework is based on a formalization of Timed CSP in the Isabelle/HOL theorem prover, we discuss formalization efforts of process algebras in interactive theorem provers in Section 3.2. Furthermore, our framework is particularly well-suited for the verification of parameterized real-time system. Therefore, we discuss different approaches for the (mostly automatic) verification of parameterized systems in Section 3.3. Finally, in Section 3.4, we give a summary of this chapter.

Related Work

3.1 Formal Verification of Real-Time Systems

In this section, we discuss approaches for the (automatic) verification of real-time systems. Most of them are based on timed automata constructions. To verify processes described in a timed process calculus, we discuss approaches for translating them to automatically analyzable languages. In the second part of this section, we discuss logics for describing (timed) properties of timed labeled transition systems, as we consider the operational semantics of Timed CSP in this thesis.

3.1.1 Model Checking Timed Systems

There exist plenty of model checking tools for the analysis of timed automata such as [DOTY95, LL98, WH02, BLN03, WWH05]. The probably most successful model checker for (networks of) timed automata is UPPAAL [BY04]. It is based on the abstraction techniques of region and zone graphs abstracting the infinite semantical state-space to a finite one and thus allows for usual model checking techniques to be applied. However, unlike timed process algebras, (networks of) timed automata do not directly allow for the convenient compositional description of real-time systems like timed process algebras. Furthermore, they are restricted to finite-state descriptions of systems.

The PAT tool [SLDP09] allows for the description of timed systems in a dialect of Timed CSP. It also contains a model checker for timed systems, which is based on the zone graph abstraction. However, currently only untimed properties can be verified for timed systems. It allows for LTL properties to be checked but without considering any quantitative timing constraints. Furthermore, it allows for refinement checking but timed edges are ignored.

3.1 Formal Verification of Real-Time Systems

The ProB [LF08] tool suite enables, among others, the verification of CSP processes. The particularity is that, in contrast to FDR2 [GRA05], LTL model checking is supported. However, verification of Timed CSP models is not supported.

Despite the lack of direct automatic verification techniques for timed process algebras, it is possible to translate a subset of these to automatically analyzable languages. Some of the existing translation-based approaches are discussed in the following. In Chapter 6, we describe the integration of automatic verification tools into our framework, which we achieve by adapting and extending the approaches of [Oua01] and [DHQ⁺08].

3.1.2 Discretization-based Analysis of Timed CSP

The motivation of [Oua01] is to relate the continuous-time semantics and a discrete-time semantics of Timed CSP in order to allow for the verification of continuous-time systems in terms of the verification of the discretely timed counterpart. The considered continuous-time semantics is the denotational timed failures semantics of Timed CSP. Using a syntactic mapping from Timed CSP to tock CSP, the FDR2 refinement checker [GRA05] can be used to verify refinement with respect to timed failures to a certain degree. In the tau-priority model of FDR2, internal steps can be given a higher priority than *tock* steps. This model is particularly necessary to model urgency of internal events in the discrete setting of tock CSP. The transformation given in [Oua01] relies on another semantical treatment of termination of parallel processes. Therefore, we adapt the transformation rules in Section 6.2 in order to treat the original semantics of parallel composition as used by FDR2. Furthermore, we give (restricted) transformation rules for *Interrupt* and *Timed Interrupt*. Still, the general translation of *External Choice* poses difficulties as discussed in Section 6.2.3. Therefore, we need to restrict the allowed processes to be transformed. A discretely

Related Work

timed version of *External Choice* will be included in some future release of FDR2 to overcome these problems. Furthermore, the FDR2 refinement checker currently only supports traces refinement in its tau-priority model. In the future, this shall be extended to the other models as well [Ros10].

3.1.3 Analyzing Timed CSP using Timed Automata Models

In [OW03], it is shown that closed timed ε-automata and a modified version of Timed CSP have equal expressive power. In the proof of this theorem, a construction of a timed automaton from a Timed CSP process is presented. Unfortunately, this construction is based on the transition system underlying the Timed CSP process and does not provide a compositional and syntactic transformation of a Timed CSP process to its timed automata counterpart. In this thesis, we prefer a syntactical transformation as the efficiency of the transformation is thereby increased. In [DHQ$^+$08], a transformation from Timed CSP to timed automata is presented, which largely fits our needs to apply UPPAAL for the automatic verification of instances of parameterized real-time systems. However, as described in Section 6.1, the proposed transformation contains flaws and is incomplete. Therefore, we give corrected transformation rules and include transformation rules for missing process operators as well.

3.1.4 Further Approaches for the Analysis of Timed CSP

Another transformation approach is presented in [DHSZ06], where the operational and timed denotational semantics of Timed CSP are encoded in a constraint-logic programming language. The semantical rules are defined

3.1 Formal Verification of Real-Time Systems

by logical rules and facts within the constraint system. In general, there is no guarantee that the proof trees are finite. Therefore, the analyzable properties are very restricted. In fact, the authors concentrate on relatively simple reachability properties and some safety properties based on traces (where the time stamps are not recorded). The only example concerning timed properties, which the authors showed for a train crossing example, is the lower bound for the occurrence of an event sequence. Additionally, this property is considered only with respect to the very initial state of the system.

In [CK05], a process algebra for timed automata based on CSP is developed. The idea is to equip CSP processes with clocks. The semantical interpretation of a process is given in terms of a timed automaton, where the reachable CSP processes are the locations of the underlying automaton. Denotational semantics in terms of region traces and region failures are developed based on the region graphs of the underlying timed automata. The denotational semantics are shown to conform to the operational timed automata semantics. Refinement is defined for the denotational region semantics and it is claimed that refinement can be verified using a CSP representation of the region automata in FDR2. However, this issue is not studied in detail. Furthermore, it is not studied how the relationship of the proposed extension is related to the semantics of Timed CSP.

Timed Systems may often be interpreted as timed labeled transition system. In our framework, we extend Hennessy-Milner logic for being able to express properties of Timed CSP processes interpreted operationally. In the following, we discuss some important logics to describe properties over labeled transition systems.

Related Work

3.1.5 Timed Modal Logics

In [Bri93], a timed modal μ-calculus is developed and examined with respect to expressibility of general properties of timed labeled transition systems. Only a strong logic is developed, which is not closed under weak (timed) bisimulation. In order to be able to consider an abstract version of the concrete parameterized real-time system within our framework, we employ weak timed bisimulation because it allows for hiding internal behavior to a certain degree.

In [LLW95], a timed modal μ-calculus with clocks for characterizing timed automata up to strong bisimilarity is developed. It is, for example, used in [LL98] to provide model checking for timed automata. However, the satisfaction of formulae in their logic is not preserved by weak timed bisimilarity, which we take as a weaker notion of semantical equivalence within this thesis.

In [MT01], the authors develop a timed version of CCS [Mil89], which was also presented in an earlier paper [MT90]. Additionally, they present a timed extension of Hennessy-Milner logic and claim that it characterizes strong (timed) bisimilarity. The authors note that this logic could be adapted for weak timed bisimilarity. The main extension that is proposed is to include a timed modality operator $\langle t \rangle \phi$. Its meaning is that there must be a path on which t time units pass such that afterwards formula ϕ is satisfied. To express properties like "event a is possible after at most t time units", infinite conjunctions could be introduced in the logic. As this would make the formalization of such a logic in a theorem prover more complex, we decided for another way by extending the untimed Hennessy-Milner logic with a general and convenient modal operator as explained in Section 5.1.3.

The presented (automatic) verification techniques either do not cope with Timed CSP processes directly or do only provide incomplete and to some degree flawed transformations to automatically analyzable languages. Therefore, we adapt and extend the transformation approaches of [Oua01] and [DHQ+08] in Chapter 6. The presented logics for describing properties of timed labeled transition systems are either too strong to be closed under weak timed bisimulation or are not well-suited to be conveniently formalized in a theorem prover. Therefore, we provide a small timed Hennessy-Milner logic in Section 4.4 to overcome these problems.

3.2 Formalization of Process Algebras in Theorem Provers

Our verification framework is based on a formalization of Timed CSP in the Isabelle/HOL theorem prover. There exists a lot of approaches concerning the mechanization of (untimed) process algebra in theorem provers. Furthermore, there are few attempts to mechanize timed process algebras. In the following, we give an overview of existing approaches.

3.2.1 IMPS (Interactive Mathematical Proof System)

In [Tha95], a formalization of timed and untimed semantical models of CSP in the IMPS theorem prover [FGT90] is presented. The particularity is that the models are defined in terms of free monoids. Unfortunately, the models are not related to the usual process operators of (Timed) CSP. Furthermore, the authors do neither describe whether this system was used for the verification of example processes nor do they give an evidence that it is even (conveniently) possible within their model.

Related Work

3.2.2 PVS (Prototype Verification System)

In [DS97], the denotational traces semantics of CSP is formalized in the PVS theorem prover [ORS92] and provided with useful proof rules for verifying security protocols. This work is extended in [ES00] by introducing event-based time in terms of a *tock* event. However, time is introduced in an ad-hoc fashion (thus, timestops are not prevented by construction) and it can be reasoned about discrete time only.

In [BH99], the authors investigate how an ACP-like process algebra can be encoded in the PVS theorem prover. They consider equational reasoning in verification. Approaches to verify meta-theoretical results as well as to verify concrete systems correct are presented. The considered process algebra is very minimalistic and only equational reasoning is considered, where equations are introduced as axioms. Therefore, complex systems cannot be described conveniently and verification can only be carried out with respect to equality of (untimed) processes.

In [WH05], the stable failures model of CSP is embedded in PVS. It is an extension of [DS97]. The formalization is used to show determinism and deadlock-freedom for an asymmetric version of the dining philosophers with arbitrarily many philosophers and deadlock-freedom of a "virtual network". Furthermore, the FDR2 refinement checker is integrated such that finite refinement proofs within a general proof in PVS can be carried out automatically. As being based on the stable failures model of CSP, the approach does not cope with reasoning about real-time processes.

In [Bro99], a partial formalization of Timed CSP in the PVS theorem prover is provided. The untimed traces, untimed failures, timed traces and timed failures semantical models are encoded. However, the semantical functions, relating process terms to semantical models are only given for some simple process constructors. As a consequence, formalization can

3.2 Formalization of Process Algebras in Theorem Provers

serve as a basis for future complete formalizations of denotational models of Timed CSP but cannot serve for performing correctness proofs in its current version.

A formalization of Timed Circus in PVS is given in [WWB10]. Timed Circus is a combination of Timed CSP and Z [WD96]. A timed denotational semantics of Timed Circus is formalized using the *Unifying Theories of Programming* [HH98]. However, the presented semantical model of Timed Circus is basically a timed traces semantics and allows "strange" processes to be defined. Furthermore, it is left for future work to apply this formalization to the verification of examples.

3.2.3 HOL System

In [Cam91], a reduced subset of CSP is formalized in the HOL theorem prover [GM93] and given a semantics in terms of the failures-divergences model. The mechanization is used to verify some of the algebraic laws of CSP. However, it is not applied to the verification of concrete systems and it does not cope with real-time systems.

In [Nes92, Nes99], the process algebra CCS is formalized in HOL. The operational semantics is inductively defined for the CCS operators and behavioral equivalences are defined in terms of weak bisimulation. Furthermore, a variant of Hennessy-Milner logic is formalized, which is closed under strong bisimulations. Using the formalization, the correctness of a buffer and a vending machine was mechanically verified. The drawback of this work is that HML properties are not preserved by weak bisimulation. Furthermore, timed systems are not considered.

Related Work

3.2.4 Isabelle

An embedding of a fair variant of (untimed) CCS in the Isabelle/HOL theorem prover [NPW02] is presented in [Com05]. The author defines strong and weak fairness of (infinite) runs of CCS processes. The formalization was used to show some meta-theoretical results but was not applied to concrete examples.

A formalization of the denotational failures-divergences semantics of CSP in Isabelle/HOL is presented in [TW97]. The authors showed that there was a flaw in the semantical treatment of terminating processes and propose a corrected semantics. The formalization is applied for showing the correctness of a simple copy process. Again, the formalization only copes with the untimed models of CSP.

An extension of [TW97] is described in [Int02]. There, a discrete-time version of the timed failures semantics of Timed CSP is considered. Like in [Bro99], the semantical model is encoded but the semantical function, giving process terms their denotational semantics is not yet defined. Besides being an incomplete formalization, we believe that the existing formalization would need fundamental redefinitions in order to interpret Timed CSP in a real-time semantics and thereby to give a sound denotational model for real-time systems.

Another formalization of CSP in Isabelle/HOL is described in [IR05]. It gives a verification environment for CSP processes where refinement proofs can be performed mechanically with respect to traces or stable failures in the untimed models of CSP.

In [RE99], weak bisimulation equivalence was formalized for labeled transition systems. The formalization was used to verify observational equivalences of different abstract levels of different protocols. However, the (untimed) protocols are described in an ad-hoc fashion as labeled transition systems.

In [RH03], the π-calculus is formalized in the Isabelle/HOL theorem prover. The main challenge here is the handling of bound names, which is needed since processes are identified up to alpha-congruence. The renaming of bound variables into fresh ones is needed when name clashes occur. This can make mechanical proofs quite tedious. Another Isabelle/HOL formalization of the π-calculus is presented in [BP07]. There, nominal logic [Urb08] is used to ease the handling of bound names. Additionally, strong and weak bisimulations are formalized. Both, the works of [RH03] and [BP07] aim at verifying meta-theoretical results and do not focus on the verification of concrete system specifications.

None of the presented approaches fully formalize Timed CSP or enable the verification of real-time systems, as we have done in this thesis. We have based our formalization on an operational semantics, not a denotational one, to exploit coalgebraic verification techniques such as bisimulation and coalgebraic invariants.

3.3 Formal Verification of Parameterized Systems

Our framework is particularly well-suited for the mechanical verification of parameterized real-time systems. In the following, we summarize the main approaches concerning the (mainly automatic) verification of parameterized systems and explain why they are not sufficient in our context.

Related Work

3.3.1 Decidable Subclasses

Because the verification problem of parameterized systems is in general undecidable [AK86], either decidable subclasses have to be identified or sound but incomplete techniques have to be provided for the (automatic) verification of parameterized systems. Decidable subclasses have, for example, been identified in [GS92], [EN95], and [EK00]. In [EN95], a class of ring-like systems is considered for which comprehensive verification can be performed by verifying finite systems of a computable *cut-off* size. In [EK00], more general but still relatively restricted subclasses of parameterized systems are considered, which make the verification problem for restricted formulae decidable. The comprehensive verification of an (untimed) parameterized system is again carried out by verifying instances up to a computable *cut-off* size.

In [AJ98], parameterized networks of restricted identical timed automata using at most one clock with an optional (untimed) control automaton are considered. Especially, location invariants are excluded, so urgency cannot be expressed in the considered model. Verification of general safety properties is decidable for this class, as illustrated with Fischer's mutual exclusion protocol. However, neither control process nor network processes may depend on the network size. This work was extended in [ADM04] where it is shown that these kinds of networks are not decidable with respect to safety properties anymore when a network process operates on two or more clocks. The authors also examine the case of discrete valued clocks and closed timed networks where this becomes decidable again.

3.3.2 Regular Model Checking

The technique of regular model checking was introduced in [KMM+97] and extended, for example, in [BJNT00] and [BLW05]. The system states

3.3 Formal Verification of Parameterized Systems

and the transition relations are represented by regular languages. To describe parameterized systems, a word of length n accepted by the automaton representing a system state is interpreted to denote the locations where each of the n network processes resides. Processes are organized in a linear, ring-like, or tree-like topology. Regular languages describing system states are manipulated by state transducers describing the transitions of the system. By computing the regular language (if it exists) that describes the set of reachable states by applying the state transducer to the initial state arbitrarily often, it can for example be checked whether the system can reach a certain set of "bad" states. The main challenge in the area of regular model checking is to provide semi-algorithms that calculate, for example, the regular language corresponding to the iterated application of the transducer to the initial locations of the system. In general, however, these algorithms are not complete. Besides the relatively unnatural description of parameterized systems in this framework, it is not applicable to real-time systems because the timing behavior cannot be described using regular languages.

3.3.3 Abstraction Techniques

Network Invariants

The abstraction technique of network invariants [WL89, KM95] was developed as a general method for verifying (linear) parameterized systems by abstracting all networks beginning from a certain size to a single invariant process. It applies an induction method for process networks where it is shown that a network of a certain fixed size refines the invariant and that the combination of the invariant process with a further process again refines the invariant. This means that the invariant represents all possible networks beginning with a certain size. If the invariant is finite, model checking can be performed on the invariant in order to show that all instances of the

Related Work

network satisfy the verified properties. In [LZDB08], the technique of network invariants is applied to the verification of self-stabilizing embedded systems. Before applying network invariants, an abstraction ensures that this technique is in principle applicable (if a network invariant can be found). In [GL08], the technique of network invariants is used in the context of timed systems (based on a dialect of timed automata) to verify a simplified version of Fischer's protocol.

The disadvantage of network invariants is that they need to be finite-state in order to be finally able to verify the abstracted system automatically. Furthermore, a finite abstraction needs to contain all system behaviors of an arbitrarily large network. Therefore, the topology of considered systems is restricted, especially if the network processes can communicate with each other and the topology changes with increasing size of the network.

Data Independence

Data-independence was studied for several formalisms and adopted for the CSP process calculus in [Laz99]. The approach focuses on systems that work on data in a restricted way. With respect to an abstract data-independent type T, processes may only input, output, nondeterministically choose values of type T, and do equality tests between values of T. Then, it can be shown that there is a finite threshold size of type T such that T can be replaced by a finite type S of this threshold size. If a refinement check with respect to S holds, then the refinement holds for all possible sizes of T. To ensure that this property holds, replicated parallel composition is not allowed over T. In [Cre01], the technique of data-independence is combined with network invariants. The main motivation is that induction alone is not sufficient when the network topology changes with increasing network sizes. By combining induction and data-independence results it

3.3 Formal Verification of Parameterized Systems

becomes however possible to handle changing communication capabilities. The essential idea of data-independent induction is to set up proof obligations for the base case and step case and to apply data-independence results to verify these proof obligations. The main goal of data-independent induction is to show that a network, which is constructed by a replicated parallel composition over type T, refines a sequential specification, which is data-independent with respect to T for all possible sizes of type T. However, this approach has not been applied in the context of Timed CSP and is limited because network processes are restricted to be data-independent and thus, they may not depend on the network size.

Counter Abstraction

In [PXZ02], counter abstraction is introduced to verify (especially liveness properties of) parameterized systems built of identical finite-state processes. The basic idea is to keep track of the number of network processes that reside in a particular state (of their finite transition system consisting of l different locations). If more than one process resides in a particular state, the exact number is abstracted by taking $k_l = 2$ to express that 2 or more processes reside in that state. If the overall system is in a state with some $k_l = 2$, and one of these processes leaves l, then either one process is in this state ore still more than 2 afterwards. Thereby, a sound (and finite) abstraction is achieved such that positive verification results can be obtained automatically and transferred to the parameterized system. In [SLR$^+$09], process counter abstraction is employed to verify parameterized systems in a fair model checking environment. Behaviorally similar processes are again grouped together and only the number of processes residing in a particular state are considered in the abstraction. In [ML09], counter abstraction is considered in the context of CSP and FDR2. It considers systems where an arbitrarily large network of processes is placed in parallel with some controller process. However, the controller process and the network

Related Work

processes are not allowed to possess process identifiers and neither network processes nor the control process may depend on the network size.

In the context of parameterized real-time systems, counter abstraction is not generally applicable because counting the processes residing in a particular state would need to abstract from clocks or timers in the system. Another disadvantage of counter abstraction is that process identifiers need to be abstracted away.

Further Approaches

In [PRZ01], the verification technique of invisible invariants is introduced to verify safety properties of restricted classes of parameterized systems. The approach is based on decidability results of certain verification conditions that occur in the verification of inductive invariants. These verification conditions can be verified for all instances of the parameterized system if they are valid for instances up to a computable *cut-off* size. By additionally providing heuristics to compute auxiliary invariants automatically, an incomplete but fully automatic verification technique is established. This work is extended in [FPPZ04] for verifying a restricted class of liveness properties for parameterized systems. The approach, however, does not cope with real-time systems.

In [KM07], data abstraction is combined with real-time systems for automatic verification using SMT solvers. Networks of timed automata are abstracted to so-called predicate diagrams for which sufficient criteria are presented to show the abstraction correct. The resulting predicate diagram is a finite-state abstraction and verification is carried out using an SMT solver or in special cases a model checker. By the use of first-order predicates in predicate diagrams, the approach also copes with parameterized real-time systems, which is shown for the verification of Fischer's protocol with arbitrarily many processes with respect to mutual exclusion. Currently, quantitative reasoning about timing behavior using this approach is not directly possible because timed edges are abstracted away in predicate diagrams. The amount of time can only be implicitly deduced from the possible valuations of the clocks in the source and the target state.

3.4 Summary

In this chapter, we have given an overview over approaches for the automatic verification of real-time systems, the mechanization of process algebras in theorem provers, and the verification of parameterized systems. The automatic verification of real-time systems is limited to finite state systems. This restriction is not necessary for the verification of process algebraic specifications in a theorem prover. However, most of the existing approaches do not cope with real-time process calculi. The (automatic) verification approaches for parameterized systems primarily focus on untimed systems and the few approaches for real-time systems are not applicable to general parameterized real-time systems. They are especially not applicable to scheduler-like systems because of their restrictions concerning the involved processes.

In the following chapters, we introduce our framework for the mechanical verification of (possibly infinite) parameterized real-time systems and its realization using a formalization of the real-time process calculus Timed CSP and transformations of (finite) instances of it to automatically analyzable languages. It copes with a large class of parameterized real-time systems such that especially scheduler-like systems can be handled.

4 A Framework for the Mechanical Verification of Parameterized Real-Time Systems

In this chapter, we present our framework for the mechanical verification of parameterized real-time systems. A former version of it is presented in [GG10b]. Our framework especially copes with parameterized systems where a distinguished control process manages a network built of arbitrarily many processes. Both, the control process and the network processes may depend on the size of the overall network. This means that the common structure of parameterized real-time systems that we consider is $N_n \stackrel{def}{=} C_n \otimes_0 (P_{1,n} \otimes P_{2,n} \otimes \cdots \otimes P_{n,n})$. In this context, $\otimes_0$ is some kind of parallel composition by which the control process C_n communicates with the network. The operator $\otimes$, also being some kind of parallel composition, is used for building up the network where, for example, messages can be exchanged between network processes. The verification goal is to show that for all possible system parameters n, N_n satisfies the given (pa-

Mechanical Verification of Parameterized Real-Time Systems

rameterized) requirements R_n that are formulated in our timed extension of Hennessy-Milner logic (HML).

The aim of our framework is to assist the design and verification of such parameterized real-time systems. It defines a verification flow, which consists of a modeling phase employing the Timed CSP process calculus, an automatic validation phase employing the UPPAAL tool suite and the FDR2 refinement checker, and an interactive verification phase employing the Isabelle/HOL theorem prover. The framework is shown in Figure 4.1 where the main design and verification flow can be followed by the numbering.

① The designer begins with a formal description of a concrete parameterized real-time system N_n with an arbitrary number of processes. This system is formalized in our mechanized Timed CSP theory in Isabelle/HOL. The designer additionally provides correctness properties R_n, which *all* instances of the parameterized system should satisfy. These properties are expressed in our timed extension of HML and formalized using our mechanization. Additionally, the designer develops an abstract parameterized system model S_n that is assumed to be (weak timed) bisimilar to the original parameterized system. This is formally shown subsequently in the comprehensive verification phase. The aim of the abstract model is to simplify the verification of the given logical requirements by exploiting its simplified structure. The overall verification goal is to show that the concrete parameterized real-time system invariantly satisfies the requirements for all network sizes after possibly hiding some events A.

② In the first validation phase, the designer uses our transformation engine *TCSP2TA* from Timed CSP to timed automata in order to transform instances of the system ($N_k \setminus A$ and $S_k \setminus A$) for a fixed network size k to timed automata. Thereby, the instances can be simulated and checked to fulfill the given requirements using, for example, the UPPAAL model checker. The aim of this phase is to find and debug potential errors with respect to the logical specification R_k in an early design phase of the parameterized system.

③ The second validation phase is used to check semantical equivalence (informally denoted by "$\approx$" in the Figure 4.1) of system instances N_k and S_k for a fixed network size k. To this end, the designer transforms instances of the parameterized Timed CSP models to their respective discretely timed counterpart in a dialect of (discretely timed) CSP, called tock CSP, using our implemented transformation engine *TCSP2tockCSP*. The semantical equivalence of these instances with respect of the timed traces model is then verified using the FDR2 refinement checker.

If the checks of one of the two validation phases are not successful, counterexamples are generated by either UPPAAL or FDR2. These can be used to debug the original parameterized system given in Timed CSP. If these checks, on the other hand, are successful then there is a good evidence that the parameterized system behaves correctly for all possible system instances. This is formally shown in the following comprehensive verification phase.

④ The first verification phase consists of proving semantical equivalence of all system instances N_i and S_i using the proof technique of weak timed bisimulations. To this end, the designer provides a witness bisimulation relation, formalizes it in Isabelle/HOL and then mechanically verifies the common proof obligations of bisimulations.

⑤ The second verification phase consists of mechanically proving that the abstract parameterized real-time system $S_n \setminus A$ invariantly satis-

Mechanical Verification of Parameterized Real-Time Systems

fies the given requirements R_n. To prove (invariant) validity of the requirements, the designer provides, similar to the case of bisimulation-based verification, a witness invariant set that contains the initial state of the system. Then, for each of the states of the invariant, the local validity of the respective formulae is verified. Furthermore, it has to be shown that each transition from a process in the invariant set again reaches a process that is contained in the invariant set. As the invariant satisfaction of formulae is preserved by weak timed bisimulations, this leads to the conclusion that also the original model $N_n \setminus A$ invariantly satisfies the respective requirements for all parameters n. Thus, the final verification goal of applying our framework is achieved.

The main advantages of our verification framework are that equivalence-oriented as well as property-oriented correctness results can be established mechanically. This ensures that possibly critical corner cases of the involved parameterized real-time systems cannot be overlooked and that parts of the formal proof can be partly automatized because by using automatized tactics in the Isabelle/HOL theorem prover. As mechanical verification in an interactive theorem prover like Isabelle/HOL is, however, relatively time-consuming, we include automatic verification tools in our framework. By employing transformation engines from (finite instances) of Timed CSP models to timed automata and tock CSP, the well developed verification tools UPPAAL and FDR2 can be applied for simulation and (instance) verification early in the design of a particular parameterized real-time system. Generated counterexamples can be used for debugging of the parameterized system.

In the following sections, we discuss these respective steps within our framework in more detail by answering the following questions.

4.1 Modeling Parameterized Real-Time Systems

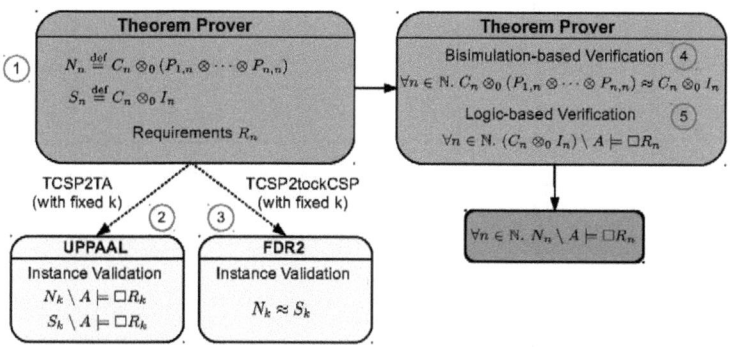

Figure 4.1: Conceptual Overview of Our Framework

Section 4.1: How are parameterized real-time systems modeled in Timed CSP?

Section 4.2: How are system instances of parameterized real-time systems validated using automatic transformation and verification tools?

Section 4.3: How are bisimulations used to show the semantical equivalence between parameterized real-time systems?

Section 4.4: How is our timed extension of HML defined and used to express and verify crucial properties of parameterized real-time systems?

4.1 Modeling Parameterized Real-Time Systems

In our framework, two Timed CSP models describing (weak timed) bisimilar parameterized real-time systems are considered. The first model describes the system in terms of arbitrarily many processes composed in parallel. To ease verification, we propose to reduce the overall complexity by developing an abstract model, which hides the complexity of the parameterization to a certain degree.

4.1.1 Parameterized Systems Composed of Parallel Processes

Parameterized real-time systems are designed to operate correctly for arbitrarily many components. For example, an operating system is, principally, able to manage arbitrarily many threads or a bus system is designed to exchange data between arbitrarily many slaves and masters connected to the bus. The parameter of these kinds of systems is therefore the number of processes. These systems can be abstractly described by a distinguished control process and a network of arbitrarily many processes. Thus, we focus on systems that can be described in the following form.

$$N_n \stackrel{def}{=} C_n \otimes_0 (P_{1,n} \otimes P_{2,n} \otimes \cdots \otimes P_{n,n})$$

For example, in an operating-system scheduler system, process C_n would be the scheduler itself while processes $P_{i,n}$ are the threads, which are controlled by it. In a bus system, C_n would consist of processes controlling the access to the bus of the masters and slaves, which are given by the processes $P_{i,n}$. To describe these kinds of systems in Timed CSP, the operator $\otimes_0$ can be expressed in terms of a parallel composition operator, while the network can be described using a replicated form of parallel composition.

Systems composed of arbitrarily many parallel processes like that above are hard to handle in (mechanical) verification. One source of complexity stems from the parameterization of the system and the thereby unbounded number of parallel processes in the network. Another source of complexity

4.1 Modeling Parameterized Real-Time Systems

stems from the timing information in the system. Therefore, we propose to simplify such kinds of parameterized systems by modeling them in a more abstract fashion. If, for example, the control process ensures that only a bounded number of timers need to be considered in each system state, then the information distributed over the individual processes of the considered process network can be recorded in process variables. Furthermore, the network of arbitrarily many processes can be replaced by an abstract process consisting of a finite amount of parallel compositions for all system parameters n. The timing properties can finally be verified on the (simplified) abstract model in order to derive them for the original concrete parameterized system. Thus, a separation between the verification concerning the parameterization and the verification concerning the properties is achieved. In the following subsection, we describe in more detail how the mechanisms of Timed CSP can be used to describe these more abstract system models.

4.1.2 Reducing Semantical Complexity of Unbounded Parallel Compositions in Parameterized Systems

To model abstract parameterized real-time systems in order to reduce the semantical complexity of parallel composition, we can make use of the (Timed) CSP mechanisms in terms of "process variables with structure". Consider, for example, the counter process as introduced in Section 2.2.

$C_0 = succ \rightarrow C_1$

$C_n = pred \rightarrow C_{n-1} \square succ \rightarrow C_{n+1}$ for $n > 0$

Here, the process variable C_i can be seen as being structured with a natural number i. The natural number keeps track of the communicated *succ* events for which no corresponding *pred* event was communicated. Thus, infinitely many process variables are used to specify the counter process.

Mechanical Verification of Parameterized Real-Time Systems

When modeling abstract parameterized systems, this mechanism can be used, for example, to keep track of the size of the process network or to record necessary distributed information of the network in process variables.

As an example, consider the following (untimed) parameterized system.

$$C \underset{\{a,b,c\}}{\|} (\|_{\{c\}\, i=1}^{n} P) \text{ where } C = (a \to C_1) \square (c \to C)$$
$$\text{and } C_1 = (a \to b \to C_1) \square (b \to C)$$
$$\text{and } P = a \to b \to c \to P.$$

The control process C manages a network of arbitrarily many processes P. Process C gives control, one after another, to one of the ready processes P by communicating the event a. However, C ensures that at most two processes are started without having communicated event b, yet. When all processes P are in the state $c \to P$, the event c is communicated, which brings the system back into its initial state where this procedure is started again.

This system can be equivalently expressed as follows (with C as above).

$$C \underset{\{a,b,c\}}{\|} I_{(n,0,n)} \text{ where } I_{(ac,bc,n)} = (ac = 0 \wedge bc = 0) \mathbin{\&} c \to I_{(n,0,n)}$$
$$\square\, ac > 0 \mathbin{\&} a \to I_{(ac-1,bc+1,n)}$$
$$\square\, bc > 0 \mathbin{\&} b \to I_{(ac,bc-1,n)}$$

Here, the process I is used to keep track of the number of processes P that can be started (ac), the number of processes that need to perform a b event after being started (bc), and the number of the processes in the network (n). When every process was started and finished its execution by communicating the b event, only the communication of c is enabled, which sets the system back into its initial state. Note that due to the presence of the control process C, we could additionally make explicit that the variable bc can only take the values 0, 1, or 2.

Already in this relatively simple (untimed) CSP model it becomes clear that the structure in the abstract model gets simpler because the distributed information of the original network is made explicit. Thereby, the abstract system does not need to be modeled using a network of (unboundedly many) parallel processes.

In this section, we have shown how parameterized real-time systems can be modeled in Timed CSP. We especially have looked at the possibility of modeling a concrete parameterized system where the network of arbitrarily many processes is directly modeled as a parallel system. Due to the semantical complexity of parallel composition, these systems are not always convenient for subsequent verification phases. Therefore, we aim at simplifying this complexity by expressing the arbitrarily large network, for example, in terms of an abstract network, which only implicitly contains the parallelism of the original system.

Before performing the comprehensive verification in Isabelle/HOL, we propose to analyze instances of the parameterized real-timed system using automatic verification tools. To this end, we have adapted, extended, and implemented transformation approaches from Timed CSP to timed automata and to tock CSP. Their integration in our framework is sketched in the following section.

4.2 Validation and Debugging of System Instances

We employ automatic verification tools in our framework to simulate and verify instances of parameterized real-time systems early in the design flow. Generated counterexamples can be used to debug the parameterized system models. If all checks succeed, the corresponding validation phase gives a good evidence that the parameterized real-time systems behave correctly for all network sizes. This is important because the relatively time-consuming comprehensive verification of the parameterized real-time system in the Isabelle/HOL theorem prover is not performed until most of the errors have been identified and corrected. Furthermore, by simulating the instance models, the gained knowledge can be used in the comprehensive verification phase. The automatic verification tools are used, on the one hand, to show that instances of the parameterized real-time system fulfill the given requirements. On the other hand, they are used to show the semantical equivalence between instances of the concrete and the abstract parameterized real-time system. Thus, both phases of the comprehensive verification are prepared and are largely automatized. To be able to employ the automatic verification tools UPPAAL and FDR2, we provide transformation engines from Timed CSP to timed automata and to tock CSP, which are based on [DHQ$^+$08] and [Oua01], respectively.

4.2.1 Transformation to Timed Automata

We extended and corrected the transformation rules given in [DHQ$^+$08], which map (finite) Timed CSP processes to a timed automata. Some of the original transformation rules concerning for example, external choice and timeout contain subtle flaws, which we corrected in adapted transformation rules. We additionally enable the transformation of *Interrupt* and *Hiding*, which are not supported in the set of the original rules. Furthermore, we have implemented our transformation rules within a Master's thesis [Wu10]. By using our transformation engine, we are able to employ the UPPAAL tool suite simulation, model checking, and debugging. This is very helpful in early design phases within our verification framework: Before performing the comprehensive and time-consuming task of mechanical verification in the Isabelle/HOL theorem prover, we translate

4.2 Validation and Debugging of System Instances

instances $N_k \setminus A$ and $S_k \setminus A$ of the parameterized real-time systems to UP-PAAL and show that the requirements R_k are satisfied for these models. In Section 6.1, we present our extended and corrected set of transformation rules from Timed CSP to timed automata. Our mapping especially copes with the *Hiding* operator of our adapted semantics of Timed CSP described in Section 5.2. It makes hidden (and thus urgent) events observable to a certain degree such that the internal behavior is also analyzable.

The transformation from Timed CSP to timed automata is relatively complex because of the translation of parallel composition, which is realized by a syntactic cross-product computation. Therefore, it is very advantageous to provide an abstract model of the parameterized real-time systems, which imposes an upper bound on the number of parallel compositions. Then, UPPAAL is especially useful for verifying properties on the abstract instances, while our transformation to tock CSP provides the possibility to give an evidence that the properties are also valid on concrete (instance) models. This is explained in the following subsection.

4.2.2 Transformation to Tock CSP

Our transformation engine from Timed CSP to tock CSP is based on the approach of [Oua01]. The motivation of [Oua01] is to relate the continuous-time semantics and a special discrete-time semantics of Timed CSP in order to allow for the verification of continuous-time systems in terms of the verification of their discretely timed counterparts. One particular aim is to provide Timed CSP with a mapping $\Psi :: TCSP \Rightarrow tockCSP$. The idea is to discretize the Timed CSP process in that time may not advance continuously but in discrete timed steps indicated by the communication of the special event *tock*. However, the proposed mapping handles termination of parallel processes in a different way than the semantics of (Timed) CSP as described in Section 2.2 or that is used by the FDR2 refinement checker:

Mechanical Verification of Parameterized Real-Time Systems

While in the original semantics of (Timed) CSP, parallel processes need to synchronize on termination events, [Oua01] allows a parallel process to terminate if one of the subprocesses can. This makes the mapping not always adequate for subsequent verification in the FDR2 refinement checker. To this end, we have changed the transformation rules to achieve this goal. Additionally, we have included new transformation rules for *Interrupt* and *Timed Interrupt*. Unfortunately, due to the handling of termination, the given rules only produce correct transformation results under several conditions. We implemented the transformation in a Master's thesis [Zho10] such that the transformation from Timed CSP to tock CSP can be performed automatically. Our adapted and extended rule set and the imposed restrictions are described in detail in Section 6.2. A transformed Timed CSP process can, to a certain degree, be analyzed using the FDR2 refinement checker. FDR2 currently supports only the traces refinement in its tau-priority model, which is needed to give internal (τ) steps a higher priority than timed steps.

Within our framework, we transform instances of the parameterized systems N_k and S_k to tock CSP and show their semantical equivalence by refinement checks with respect to timed traces in both directions. Thereby, we get the evidence that the bisimulation proof in the comprehensive verification phase of our framework is principally possible. If one of these checks failed, the bisimulation proof could not be established as this would imply timed traces refinement for all instances. Counterexamples generated by FDR2 can be simulated using ProB [LF08] or ProBE [PRO03]. However, as neither of these tools support the tau-priority model, τ events have to be performed manually in favor of possible tock events.

If the considered instances of the given parameterized real-time systems behave as expected as explored in UPPAAL and FDR2, there is a good evidence that the systems also behave as expected for large network sizes. However, automatic verification is always restricted to finitely many and relatively small models. Therefore, we propose to perform the comprehensive verification using our Timed CSP mechanization in the Isabelle/HOL theorem prover. To establish the overall verification goal, two verification steps are performed: bisimulation-based verification and logic-based verification. In the following section, we start with discussing the bisimulation-based verification phase.

4.3 Bisimulation-Based Verification of Parameterized Timed CSP Models

As described in Section 4.1, we assume that a designer starts with a description of the system by explicitly modeling the parallel structure of the network, which is controlled by a distinguished control process. To simplify the structure of this model, the designer develops an abstract model of the parameterized system in order to ease later verification with respect to the given requirements. These two systems are assumed to be behaviorally equivalent. This is formally and mechanically proven using the notion of weak timed bisimulation as introduced in Section 2.1.3.

The proof obligation for the semantical equivalence of these two parameterized Timed CSP models is the following.

$$\forall\, n \in \mathbb{N}.\ C_n \otimes_0 (P_{1,n} \otimes P_{2,n} \otimes \cdots \otimes P_{n,n}) \approx C_n \otimes_0 I_n$$

The equivalence can be shown by developing a weak timed bisimulation relation, which includes the process pairs of interest. In the case of such parameterized systems as above, the structure of the bisimulation relation can be given as follows.

$$B \stackrel{def}{=} B_1 \cup B_2 \cup \ldots \cup B_m$$

Mechanical Verification of Parameterized Real-Time Systems

where each of the B_i has the following structure.

$$B_i \stackrel{def}{=} \{(P,Q).\quad \langle *\text{Variables}* \rangle$$
$$\exists k\ Contr\ Procs\ I\ \ldots$$
$$\langle *\text{Parameter Conditions}* \rangle$$
$$k > 0 \wedge \ldots$$
$$\langle *\text{Involved Processes}* \rangle$$
$$Contr = \ldots \wedge$$
$$\forall i. i \geq 1 \wedge i \leq k \longrightarrow Procs\ i = \ldots \wedge$$
$$I = \ldots \wedge$$
$$\ldots$$
$$\langle *\text{Definition of the Tuple}* \rangle$$
$$P = Contr \otimes_0 (\otimes_{i=1}^{k} Procs\ i) \wedge$$
$$Q = Contr \otimes_0 I\ \}$$

By splitting the bisimulation relation into a finite number of subrelations B_i, we can keep the proof terms in the corresponding mechanical bisimulation proof syntactically small. This is very important for mechanical verification because the terms are parsed on-the-fly and therefore the terms must not be too complex. The subrelations correspond to elementary states in which the parameterized systems can possibly reside. In Chapter 7, we report on a case study where we successfully make use of this style for defining a bisimulation relation.

To prove the weak timed bisimulation relation correct, the common proof obligations for bisimulations (see Section 2.1.3) must be shown for each subrelation B_i. This means that the following must hold.

$$\forall i \in \{1..m\}.$$
$$\forall (P,Q) \in B_i.\ \forall \beta.$$
$$P \xrightarrow{\beta} P' \quad \longrightarrow \quad \exists Q'.\ \exists j \in \{1..m\}.\ Q \xrightarrow{\beta}_{wt} Q' \wedge (P',Q') \in B_j \wedge$$
$$Q \xrightarrow{\beta} Q' \quad \longrightarrow \quad \exists P'.\ \exists j \in \{1..m\}.\ P \xrightarrow{\beta}_{wt} P' \wedge (P',Q') \in B_j$$

4.3 Bisimulation-Based Verification

As explained in Chapter 5, we provide an inductive formalization of Timed CSP's operational semantics. Therefore, possible steps of a Timed CSP process can be deduced based on case distinction and answering steps can be constructed according to the inductive rules. It is easy to see that the proof schema above enables us to show that $B = B_1 \cup \ldots \cup B_m$ is indeed a bisimulation relation. Note that this schema comprises the equivalence proof for all sizes of the network. This is achieved by the parameter k in the subrelations B_i indicating some arbitrary size of the network. Thereby, B_i does not only relate system instances for a fixed k but for all valid k. In this context, "valid" means that k can be constrained to useful values. In the schema above, k is constrained to be greater than 0. However, it would also be possible to consider only networks of size at least, say, 3.

The goal of the bisimulation-based verification phase is to establish (weak timed) bisimilarity between the original and the simplified abstract parameterized real-time system. The advantage of an infinite bisimulation relation like the one above is that it establishes equivalence for all network sizes. At the same time, by splitting the bisimulation relation in subrelations, the terms are kept small enough to be processed effectively in Isabelle/HOL.

In the next section, we present the second verification phase and the last phase in the application of our framework. We formally introduce our timed extension of HML and discuss its suitability to express crucial properties of parameterized real-time systems. Furthermore, we discuss how a parameterized real-time system can be shown to satisfy given requirements.

4.4 Logic-Based Verification of Parameterized Timed CSP Models

Within our framework, the goal is to verify that the initially developed concrete parameterized real-time system fulfills safety, liveness and timing properties. We have discussed how the original system can be shown to be equivalent to a simplified abstract system description in the last section. In this section, we discuss how crucial logical properties can be shown to be satisfied by the abstract system model. To this end, we develop a timed modal logic following HML [HM80]. Then, we explain how to use this logic to express interesting properties of real-time systems. Finally, we discuss how to verify the (invariant) satisfaction of a logical formula by a (parameterized) Timed CSP model.

4.4.1 Timed Hennessy-Milner Logic

HML is a logic that is particularly useful for describing properties of processes in labeled transitions systems (LTSs). It is a simple modal logic in which it is possible to express that a certain transition is enabled and after performing this transition a certain formula must hold ($\langle \alpha \rangle F$). Furthermore, it can be expressed that after each transition of a certain form a certain formula must be true ($[\alpha]F$). See Section 2.1.4 for a deeper presentation of this logic.

We define our timed version of HML based on a timed LTS (S, T, A, D). Its syntax is given as the follow.

Definition 15 (Syntax of Timed Hennessy-Milner Logic) *Let p range over predicates on timed events. Then, the formulae of timed HML are given according to the following recursive definition.*

4.4 Logic-Based Verification

$$\phi := tt \mid \neg\phi \mid \phi_1 \wedge \phi_2 \mid \langle\langle p \rangle\rangle \phi$$

On first sight, this logic is identical to the original logic. However, the difference comes from the meaning of p. It is a predicate on timed events (t, a), where $a \in A$ and $a \neq \tau$. The formula $\langle\langle p \rangle\rangle \phi$ intuitively means that at least one timed event of p can be performed such that ϕ holds afterwards.

Definition 16 (Semantics of Timed Hennessy-Milner Logic) *The timed semantics of timed HML is recursively defined as follows.*

$$
\begin{aligned}
P &\models tt & &\text{iff} & &\text{true} \\
P &\models \neg\phi & &\text{iff} & &\neg(P \models \phi) \\
P &\models \phi_1 \wedge \phi_2 & &\text{iff} & &(P \models \phi_1) \wedge (P \models \phi_2) \\
P &\models \langle\langle p \rangle\rangle \phi & &\text{iff} & &\exists (t,a)\, P'.\ P \stackrel{(t,a)}{\approx\!\!\!>} P' \wedge p\,(t,a) \wedge P' \models \phi
\end{aligned}
$$

The most interesting case is the last one. The statement $P \models \langle\langle p \rangle\rangle \phi$ is valid if there exists a timed event (t, a) satisfying predicate p such that process P may evolve to some P' in t time units and with event a being communicated[1] and that the reached process P' satisfies formula ϕ.

To conveniently express important properties, we first define the standard abbreviations within timed HML.

$$
\begin{aligned}
\mathit{ff} &\stackrel{def}{=} \neg tt \\
\phi_1 \vee \phi_2 &\stackrel{def}{=} \neg(\neg\phi_1 \wedge \neg\phi_2) \\
\phi_1 \longrightarrow \phi_2 &\stackrel{def}{=} \neg\phi_1 \vee \phi_2 \\
[[p]]\phi &\stackrel{def}{=} \neg\langle\langle p \rangle\rangle\phi
\end{aligned}
$$

Furthermore, we derive convenient operators for describing lower and upper time bounds with respect to the occurrence of some particular event as follows.

---
[1]Time-event steps $\stackrel{(t,a)}{\approx\!\!\!>}$ are introduced in Section 2.1.2.

Mechanical Verification of Parameterized Real-Time Systems

$$\langle\langle a \rangle\rangle_{\sim t} \phi \stackrel{def}{=} \langle\langle \lambda(r,x). \, x = a \wedge r \sim t \rangle\rangle \phi$$
$$[[a]]_{\sim t} \phi \stackrel{def}{=} [[\lambda(r,x). \, x = a \wedge r \sim t]] \phi$$

In this context, $\sim$ is an operator from the set $\{<, \leq, =, >, \geq\}$. This means that $\langle\langle a \rangle\rangle_{\sim t}$ is used to describe that event a must be possible within the time bounds specified by $\sim t$ and that ϕ holds thereafter (after performing the respective timed event). The formula $[[a]]_{\sim t}$ on the other hand means that whenever event a is possible within the time specified by $\sim t$, ϕ must hold thereafter.

We additionally introduce a finite universal quantifier, which is actually realized as a finite conjunction of indexed formulae.

$$\bigwedge [m, n] \phi_i \stackrel{def}{=} \phi_m \wedge \phi_{m+1} \wedge \cdots \wedge \phi_{n-1} \wedge \phi_n$$

Note that ϕ is a function, which assigns a formula to a natural number. Thereby, we can make use of meta-logical expressions, for example, comparisons of i to a constant. See below for examples on using this mechanism.

Also note that all formulae that can be described in our timed adaption of HML are finite. This has the advantage that instances of parameterized formulae ϕ_n can be expressed more easily in the CTL dialect of UPPAAL, which we employ to verify instances of a parameterized system. In Section 6.1, we discuss this issue in detail.

4.4.2 Examples

As a first example, consider the following formula.

$$[[a]]_{=0} \, ([[a]]_{<d_l} f\!f \wedge [[a]]_{>d_u} f\!f)$$

4.4 Logic-Based Verification

It expresses that after the communication of an a event of a process P, the consecutive communication (if possible) of an a event is only possible in the interval $[d_l, d_u]$. In other words, two consecutive a events are separated by at least d_l and by at most d_u time units.

To express a sort of bounded response properties, the logic can be used in the following way. After an a event occurred, there is an internal path where event b is enabled thereafter and it is ensured that b may only occur within the time bounds d_l to d_u. This is expressed in the following formula.

$$[[a]]_{=0} (\langle\langle b \rangle\rangle_{\geq 0} tt \wedge [[b]]_{<d_l} f\!f \wedge [[b]]_{>d_u} f\!f)$$

As an example, especially in the context of parameterized systems, consider the following. Assume that for each network process i (ranging from 1 to n) there exists a dedicated event a_i denoting a special action of it. To express, for example, that the communication of two consecutive events a_i, a_j is only allowed if $i < j$, we can use the following formula.

$$\bigwedge[1,n]\, (\lambda i.\ [[a.i]]_{=0}\, \bigwedge[1,n](\lambda j.\ \textbf{if } j \leq i \textbf{ then } [[a.j]]_{\geq 0} f\!f \textbf{ else } tt))$$

It says that after the communication of an $a.i$ event, the communication of $a.j$ with j less or equal to i must not happen. This formula exploits that the finite universal quantifier $\bigwedge[m,n]\phi_i$ involves the function ϕ mapping natural numbers to formulae. Thereby, we can make use of the meta-logical *if-then-else* statement to compare the process identifiers i and j.

There are two important things that need to be taken into account: First, our timed HML is not designed as a temporal logic. This means that for example the formula $[[a]]_{=0} \langle\langle b \rangle\rangle_{\geq 0} tt$ can only be fulfilled by a process P if a is disabled in the initial state, or, if enabled, after performing a, event b is enabled. This especially means that this formula is not fulfilled by a process that must communicate the visible event c first before being able to communicate event b. In this case, c would need to be hidden and must not be observable. Second, when instantiating this logic in the context of

Mechanical Verification of Parameterized Real-Time Systems

Timed CSP, note that visible events cannot be enforced to happen except by our conservative extension of the operational semantics (Chapter 5) in terms of observable events, which are hidden beforehand to gain urgency.

4.4.3 Coinductive Invariants

To formally describe properties that are valid in all reachable states of a process, we use the concept of coinductive invariants. The intuitive meaning is that a process P invariantly satisfies a certain formula if it holds in its current state and that each derivative of P again invariantly satisfies the formula and so on.

Definition 17 (Coinductive Invariant) *A set of processes CI is a coinductive invariant with respect to formula ϕ if the following holds.*

$$\forall P \in CI. \ P \models \phi \land (\forall \beta \ P'. \ (P \xrightarrow{\beta} P') \longrightarrow (P' \in CI))$$

A process P *invariantly satisfies* ϕ (written as $\Box \phi$) if a coinductive invariant CI with respect to ϕ exists such that $P \in CI$.

4.4.4 Logical Verification

Based on its timed operational semantics, Timed CSP can be interpreted as a timed LTS. Therefore, we can employ our timed HML for Timed CSP processes. For the verification of the assertion $P \models \psi$ for some (possibly parameterized) Timed CSP process P and timed HML formula ψ, we can directly use the operational semantics. The only proof obligations that cannot be further simplified are $P \models \langle\langle p \rangle\rangle \phi$ and $P \models \neg \langle\langle p \rangle\rangle \phi$.

The assertion $P \models \langle\langle p \rangle\rangle \phi$ is equivalent to

$$\exists P' \ t \ a. \ P \stackrel{(t,a)}{\Rightarrow} P' \ \land \ p(t,a) \land P' \models \phi$$

4.4 Logic-Based Verification

The assertion $P \models \neg \langle\langle p \rangle\rangle \phi$ is equivalent to

$$\forall P' \; t \; a. \; P \stackrel{(t,a)}{\approx\!\!>} P' \wedge p(t,a) \longrightarrow \neg (P' \models \phi)$$

The first proof obligation can be verified by finding witnesses P', t and a such that $p(t,a)$, Then, the inductive rules of Timed CSP's operational semantics can be applied to verify that the corresponding path exists. The second proof obligation is harder in that it needs to be shown that for all possible $\stackrel{(t,a)}{\approx\!\!>}$ steps satisfying predicate p the reached process does not satisfy formula ϕ. To span the relevant tree of the operational semantics, we make use of deduction rules of time-event steps and of the operational semantics, which can partly be applied automatically. See Chapter 5 for more details on our formalization of Timed CSP.

Within our verification framework, we aim at mechanically proving that $S_n \setminus A \models \Box R_n$ for all network sizes n. To achieve this, a parameterized invariant set CI_n has to be provided, which is again split into m subsets $CI_{i,n}$ such that $CI_n = CI_{1,n} \cup \ldots \cup CI_{m,n}$. The subsets $CI_{i,n}$ correspond to elementary states in which the abstract parameterized system of size n can possibly reside. First, it has to be verified that for each network size n, the initial state of the abstract parameterized real-time system is contained in some $CI_{i,n}$.

$$\forall n \in \mathbb{N}. \; \exists j \in \{1,\ldots,m\}. \; S_n \setminus A \in CI_{j,n}$$

Furthermore, it needs to be shown that CI_n is indeed an invariant. This means that each process $P \in CI_{i,n}$ (locally) satisifies the requirements R_n and that each derivative of P is again contained in some $CI_{j,n}$. Formally, this is expressed in the following proof schema.

Mechanical Verification of Parameterized Real-Time Systems

$\forall n. \forall i \in \{1..m\}.$
$\quad \forall P \in CI_{i,n}.\ P \models R_n\ \wedge$
$\quad\quad (\forall \beta\ P'.\ P \xrightarrow{\beta} P' \longrightarrow \exists j \in \{1..m\}.\ P' \in CI_{j,n})$

Note that it is necessary to parameterize the invariant set CI by the network size. Otherwise, the assertion $P \models R_n$ in the above schema would express that process P satisfies the requirements R_n where P would describe some state of the parameterized system of an arbitrary network size. Thus, there would be no relationship between the network size of the abstract parameterized system and the parameter of the requirements.

When the proof for $\forall n \in \mathbb{N}.\ S_n \setminus A \models \Box R_n$ has been performed, then the requirements can be shown to hold for the concrete parameterized system $N_n \setminus A$ as well: First, from the weak timed bisimilarity of N_n and S_n for arbitrary n, we can conclude, due to the congruence property of weak timed bisimilarity with respect to *Hiding*, that $N_n \setminus A$ is weak timed bisimilar to $S_n \setminus A$. Second, from the preservation of the invariant satisfaction of timed HML formulae by weak timed bisimulation, we can conclude that also $N_n \setminus A \models \Box R_n$ for all network sizes. Both, the congruence property of weak timed bisimulation and the preservation of the satisfaction with respect to (invariant) formulae is verified using our mechanizations of Timed CSP, bisimulations, and timed HML in the Isabelle/HOL prover, which are presented in the next chapter.

In this section, we have presented a timed extension of HML with which crucial properties of parameterized real-time systems can be expressed. Due to the fact that this logic makes local assertions about processes of a timed LTS (for example that of Timed CSP), we have introduced invariants such that invariant satisfaction of formulae can be expressed. Similar to the case of bisimulation-based verification, we propose to split the witness set for invariants into subsets such that the term structure can be effectively handled in our Timed CSP formalization in Isabelle/HOL. The verification of local validity of formulae can be partly automatized in Isabelle/HOL based on the operational semantics.

4.5 Summary

In this chapter, we have presented the structure of our framework for the mechanical verification of parameterized real-time systems. It assumes that a designer provides a concrete model of a parameterized real-time system, an abstract model of it that eases mechanical verification, and requirements that all instances of the concrete model should satisfy. The aim of the framework is to support the mechanical verification of the concrete parameterized system using a bisimulation-based verification phase and a logic-based verification phase. Bisimulations are used to show the semantical equivalence between the concrete and an abstract model of a parameterized real-time system. To express properties of parameterized real-time systems, we have developed a timed extension of Hennessy-Milner logic and have discussed how the (invariant) satisfaction of formulae can be proved. To check the (invariant) satisfaction of the requirements on system instances prior to the relatively time-consuming task of the comprehensive verification of the parameterized system in the Isabelle/HOL theorem prover, we employ transformation engines from Timed CSP to timed automata and to tock CSP. This enables automatic simulation and verification of system instances using the UPPAAL tool suite and the FDR2 refinement checker. Possible counterexamples can be used to debug the parameterized systems. The comprehensive verification of the overall parameterized real-time system is performed in the Isabelle/HOL theorem prover. To this end, we have developed formalizations of Timed CSP, bisimulations, and of our timed version of Hennessy-Milner logic in Isabelle/HOL, which are presented in detail in the next chapter.

5 Formalization of Timed CSP in the Isabelle/HOL Theorem Prover

In this chapter, we present the foundations for the mechanical verification of our framework. We give an overview of our formalization of Timed CSP together with bisimulation- and property-based verification techniques in the Isabelle/HOL theorem prover. To achieve a general theory and a high reusability, we first formalize the notion of (timed) labeled transition systems (LTS), different kinds of bisimulations and our timed Hennessy-Milner logic (HML). The main advantage of this approach is twofold: First, all definitions and lemmas for timed LTSs can be reused for concrete models describing a timed LTS (such as Timed CSP). Second, definitions and properties that are defined on the level of timed LTSs are separated from language-specific issues, which enhances modularity and again improves reusability. Then, we interpret the operational semantics of Timed CSP as a timed LTS. By this, we instantiate bisimulations and timed HML for

Formalization of Timed CSP in the Isabelle/HOL Theorem Prover

Timed CSP. This allows us, on the one hand, to relate behaviorally equivalent Timed CSP processes and, on the other hand, to express and verify correctness requirements of processes concisely. Parts of our Isabelle/HOL formalizations have been presented in [Göt07], [GG09] and [GG10a].

This chapter is divided into four sections: In Section 5.1, we introduce a theory for timed LTSs in Isabelle/HOL and formalize different kinds of bisimulations and our timed HML. Furthermore, we present verification results for important properties on this level of abstraction, for example, that timed HML formulae are preserved under weak timed bisimulation. In Section 5.2, we present our formalization of Timed CSP and show how the verification techniques above can be instantiated in this context. Finally, in Section 5.3, we discuss special definitions and related lemmas that enable the concise description and verification of parameterized real-time systems. The chapter closes with a summery concerning our Isabelle/HOL formalizations in Section 5.4.

5.1 Fundamental Theories

In our formal verification framework, we use the operational semantics of Timed CSP, which gives a semantical basis for performing correctness proofs about (parameterized) real-time systems. An abstract view on this semantics yields a timed LTS. Important verification techniques like different kinds of bisimulations and our timed HML can be established on this level of abstraction. We therefore developed a generic formalization of these verification techniques, which can be instantiated for concrete timed LTSs like, for example, that of Timed CSP. By this, we inherit all abstract definitions and lemmas on the level of (timed) LTSs for concrete instantiations. As a consequence, our approach can also cope with other timed process algebras, as long as they can also be interpreted as timed LTSs.

5.1.1 Labeled Transition Systems

In Section 2.1, we have introduced the general notion of (timed) LTSs. It is based on a set of triples consisting of source and target states that are connected by labeled edges. In Isabelle, we represent such systems using the type abbreviation lts. This type synonym is parameterized over a type variable 's, which represents the states (also called processes), and a type variable 'a, which represents the labels of the underlying LTS.

```
type_synonym ('s,'a) lts = ('s × 'a × 's) set
```

To conveniently define LTSs that allow for the distinction between internal and external labels/timed labels, we use the structuring concept of locales in Isabelle. LTSs that explicitly consider internal (τ) steps are then defined within the locale of basic_LTS.

```
locale basic_lts =
  fixes T:: ('s,'a) lts
    and tau:: 'a        ($\tau$)
```

Here, the underlying LTS is fixed together with a special event tau, which is interpreted as an internal event and syntactically abbreviated by τ.

Timed LTSs extend basic LTSs by additionally introducing timed edges. In our formalization, we use the time domain of positive reals. However, it would be easy to adapt it for other time domains. Within the locale timed_lts we assume that the timed edges fulfill several properties. For example, a timed edge can be split into consecutive timed edges with the same overall time span and time does not introduce further non-determinism.

Formalization of Timed CSP in the Isabelle/HOL Theorem Prover

```
locale timed_lts = basic_lts +
  fixes time :: real ⇒ 'a
  assumes inj_time : inj time
    and neq_time_tau : time t ≠ τ
    and time_pos : (P, time t, P') ∈ T ⟹ t > 0
    and time_interpolation :
          ⟦ (P1, time (t1+t2), P3) ∈ T ; t1 > 0 ; t2 > 0 ⟧
          ⟹ ∃ P2. (P1, time t1, P2) ∈ T ∧
                  (P2, time t2, P3) ∈ T
    and time_determinism : ⟦ (P, time t, P1) ∈ T ∧
                             (P, time t, P2) ∈ T ⟧ ⟹ P1=P2
```

In Section 2.1.2, we have defined the extended step relations $\longrightarrow_w$ and $\longrightarrow_{wt}$, which abstract away from internal steps ($\longrightarrow_w$) and internal steps and the amount of intermediate timed steps ($\longrightarrow_{wt}$), respectively. Furthermore, we have introduced compound time-event steps $\stackrel{(t,a)}{\approx\!\gg}$. We inductively defined these extended step relations within the context of the locales above. Although these definitions are differently named in our Isabelle/HOL formalization, we use the originally introduced syntax in the rest of this chapter. On this basis, we can define, for example, strong, weak and weak timed bisimulations, which are well-suited to express and verify the conformance between different system models, as explained in the following subsection.

5.1.2 Abstract Bisimulations

The common structure of all (considered) kinds of bisimulations is that some step of one process with respect to the underlying LTS must be adequately answered by the second process (and vice versa). The second process is possibly allowed to take a "complex" step where, for example, from internal steps is abstracted to a certain degree. Thus, bisimulations are defined with respect to an original step relation T and an extended step relation $\hat{T}$, which can concisely be considered in the locale Bisimulation.

5.1 Fundamental Theories

```
locale Bisimulation =
  fixes T:: ('s,'a)lts
    and T̂:: ('s,'a)lts
  assumes ...
```

We define bisimulations abstractly for arbitrary step relations T and T̂ so that we can instantiate these definitions (and hence inherit abstract properties) for strong, weak and weak timed bisimulation in the context of basic or timed LTSs. In the context of the locale `Bisimulation`, we define bisimilarity as a coinductive set.

```
coinductive_set bisimilar::('s,'a)lts ⇒ ('s,'a)lts ⇒ ('s×'s)set
for T:: ('s,'a)lts and T̂:: ('s,'a)lts where
⟦ ∀ e P2. (P1,e,P2) ∈ T ⟶ (∃ Q2.
                (Q1,e,Q2) ∈ T̂ ∧ (P2,Q2) ∈ bisimilar ∧
   ∀ e Q2. (Q1,e,Q2) ∈ T ⟶ (∃ P2.
                (P1,e,P2) ∈ T̂ ∧ (P2,Q2) ∈ bisimilar ⟧
   ⟹ (P1,Q1) ∈ bisimilar
```

The advantage of defining bisimilarity as a coinductive set is that the following coinduction proof scheme is automatically provided by Isabelle.

$$\frac{\begin{array}{l}(P,Q) \in X \\ (\forall (P1,Q1) \in X. \\ \quad \forall e\ P2.(P1,e,P2) \in T \longrightarrow (\exists Q2.(Q1,e,Q2) \in \hat{T} \\ \qquad \land\ (P2,Q2) \in X \cup bisimilar) \\ \quad \forall e\ Q2.(Q1,e,Q2) \in T \longrightarrow (\exists P2.(P1,e,P2) \in \hat{T} \\ \qquad \land\ (P2,Q2) \in X \cup bisimilar))\end{array}}{(P,Q) \in bisimilar}$$

This yields a powerful proof principle to verify two concrete processes P and Q bisimilar. The set X can be instantiated with a concrete bisimulation relation. First, it has to be shown that the tuple (P,Q) is included in X. Then, for all tuples within X it has to be verified that each step of one process can be adequately answered by the other and that they thereby

Formalization of Timed CSP in the Isabelle/HOL Theorem Prover

reach processes that occur again in X or which are already known to be bisimilar.

On this level of abstraction, we are able to prove under weak additional assumptions that bisimilarity is an equivalence relation, i.e., it is reflexive, symmetric, and transitive. These assumptions, which hide behind "..." in locale Bisimulation, are that $T \subseteq \hat{T}$ and that also complex $\hat{T}$ steps can be answered by $\hat{T}$ steps within a bisimulation relation.

In order to define strong, weak and weak timed bisimulations, we instantiate T and $\hat{T}$ appropriately. This can conveniently be done using the sublocale mechanism of Isabelle.

```
sublocale  basic_lts ⊆ strong: Bisimulation T T

sublocale  basic_lts ⊆ weak: Bisimulation T ⟶w

sublocale  timed_lts ⊆ weak_timed: Bisimulation T ⟶wt
```

Thereby, we inherit the properties, which have been verified in the context of abstract bisimulations, i.e., all considered kinds of bisimulations are, in particular, reflexive, symmetric, and transitive relations.

Additionally, we have verified that strong bisimilar processes are also weak bisimilar and that weak bisimilar processes are also weak timed bisimilar. These properties are due to the fact that $T \subseteq \longrightarrow_w$ and that $\longrightarrow_w \subseteq \longrightarrow_{wt}$. This hierarchy of the considered kinds of bisimulations allows for taking the simpler strong or weak notion of bisimulation in order to show weak timed bisimilarity of two processes.

So far, we have introduced a generic formalization of (timed) LTSs and bisimulations, which we have instantiated with strong, weak and weak timed bisimulation on the abstraction level of (timed) LTSs. Bisimulations are very well-suited to express and verify the semantical equivalence of (timed) processes. However, they lack convenient mechanisms to express properties of processes. Therefore, we developed and formalized a timed extension of HML, which is presented in the next subsection.

5.1.3 Timed Hennessy-Milner Logic

In Section 4.4, we sketched a timed extension of HML to express crucial properties of real-time systems. Compared to the untimed version of HML, we propose to replace the original possibility modality $\langle a \rangle$ by a weak timed modality $\langle\langle p \rangle\rangle$, where p is a predicate on timed events (t, a). By this extension, we keep the logic very small but at the same time we gain enough expressiveness to describe crucial properties of real-time systems. In Isabelle/HOL, we define the syntax of this logic by the following datatype within the context of `timed_lts`.

```
datatype 'b formula =
  tt
| notF "'b formula"                              ("¬_f_")
| andF "'b formula" "'b formula"                 (infixl "∧_f")
| poss "(real × 'b ⇒ bool)" "'b formula"         ("⟨⟨_⟩⟩_")
```

Note that we give the index f to the logical operators of our logic to distinguish them from the object logic connectives of Isabelle/HOL.

The semantics is then given by the recursive function `sem`, which can be abbreviated by the convenient infix notation $\models$.

```
primrec sem:: 'a ⇒ 'b formula ⇒ bool (infix "|=")
where
  "P |= tt = True"
| "P |= (¬_f φ) = (¬ (P |= φ))"
| "P |= (φ_1 ∧_f φ_2) = ((P |= φ_1) ∧ (P |= φ_2))"
| "P |= (⟨⟨p⟩⟩ φ) = (∃ Q (t,a). P ⇒̃^{(t,a)} Q ∧ p(t,a) ∧ Q |= φ)"
```

The necessity operator for our timed HML is straightforwardly introduced by the following syntactical abbreviation.

Formalization of Timed CSP in the Isabelle/HOL Theorem Prover

```
abbreviation nec:: "(real×'b ⇒ bool) ⇒ 'b formula ⇒ 'b formula"
                                                ("[[_]] _")
where
  "[[p]] φ ≡ ¬f (⟨⟨p⟩⟩ ¬f φ)"
```

Furthermore, we define convenient abbreviations to express lower and upper time bounds for certain events using the previously introduced possibility and necessity modality. As an example consider the formula $\langle\langle a \rangle\rangle_{\leq t}$, which expresses that some process is able to perform event a after at most t time units. This is defined using the following abbreviation by instantiating the predicate p appropriately.

```
abbreviation poss_leq:: "'b ⇒ real ⇒ 'b formula ⇒ 'b formula"
                                                ("⟨⟨_⟩⟩≤_")
where
  "⟨⟨a⟩⟩≤t φ ≡ ⟨⟨ (λ (r,x) . x=a ∧ r≤t) ⟩⟩ φ"
```

In a similar fashion, we formalize the abbreviated operators $\langle\langle a \rangle\rangle_{\sim t} \phi$ and $[[a]]_{\sim t}\phi$ with $\sim \in \{<, \leq, =, \geq, >\}$.

Altogether, we have presented a timed logic for expressing properties of processes in a timed LTS. The assertion $P \models \phi$ means that formula ϕ holds in P's initial state. However, we often wish to express that a certain property holds in *all* reachable states from P's initial state. Therefore, we use the concept of coinductive invariants as explained in the following.

Coinductive Invariants

Like in the case of bisimilarity, we define invariants with respect to a timed HML formula ϕ coinductively.

```
coinductive_set invariant:: 'a formula ⇒ 's set
for φ :: 'a formula
where ⟦ P ⊨ φ ; ∀ e P'. (P,e,P') ∈ T ⟶ P' ∈ invariant φ ⟧
      ⟹ P ∈ invariant φ
```

5.1 Fundamental Theories

This definition expresses that a process P invariantly satisfies formula ϕ if it (initially) satisifies ϕ and all derivatives of P are again contained in the invariant, i.e., invariantly satisfy ϕ. Again, Isabelle automatically provides a coinduction scheme, which looks as follows.

$$\frac{P \in S \quad (\forall Q \in S.\ Q \models \phi \wedge \forall e\ Q'.\ (Q, e, Q') \in T \longrightarrow Q' \in S \cup invariant\ \phi)}{P \in invariant\ \phi}$$

Thus, in order to verify that a concrete process P invariantly satisfies formula ϕ, one has to find a set S, which contains P. Furthermore it has to be verified that all processes in S initially satisfy ϕ and that their derivatives are again in S or are already known to invariantly satisfy ϕ.

We have presented generic formalizations of bisimulations and our timed HML. Bisimulations are well-suited for showing the semantical equivalence of processes of (timed) LTSs. Our timed HML is well-suited for showing (invariant) timed properties of processes. In the following subsection, we show that the satisfaction of timed HML formulae is preserved by weak timed bisimilarity.

5.1.4 Preservation of Timed HML under Bisimulation

An important property of our timed HML is that weak timed bisimilar processes satisfy exactly the same formulae. This property is stated in the following lemma.

```
lemma weakt_bisim_presIff: "(P,Q) ∈ weak_timed.bisimilar
         ⟹ P ⊨ φ ⟷ Q ⊨ φ"
```

We have verified this property using induction over logical formulae and the additional property that for weak timed bisimilar processes $\stackrel{(t,a)}{\approx\!\!\!\Rightarrow}$ steps can be answered by $\stackrel{(t,a)}{\approx\!\!\!>}$ steps reaching again bisimilar processes. This lemma is used to show that also the invariant satisfaction of timed HML formula is preserved by weak timed bisimulation.

```
lemma weakt_bisim_invIff:  "(P,Q) ∈ weak_timed.bisimilar
                        ⟹ P ∈ invariant ϕ ⟷ Q ∈ invariant ϕ"
```

We have verified this lemma by coinduction for invariants using the witness set $\{q.\ \exists p.\ p \in \textit{invariant}\ \phi \land (p,q) \in \textit{weak\_timed.bisimilar}\}$, which is closed under the rules of the coinductive invariant definition. This property allows for the following verification principle, which our framework is based on: To show that a possibly complex process P invariantly satisfies formula ϕ (denoted as $\Box \phi$ in the figure below), we can prove this property on a simpler process Q, which is weak timed bisimilar to P.

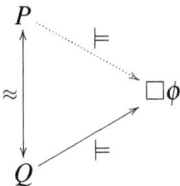

In this section, we have shown how we have defined generic bisimulations and a timed HML in Isabelle. We have instantiated these abstract bisimulations in order to define and inherit common properties of strong, weak and weak timed bisimulations. Furthermore, we have shown that two weak timed bisimilar processes invariantly satisfy the same timed HML formulae. In the next section, we present our formalization of Timed CSP in the Isabelle/HOL theorem prover and show how the presented verification techniques can be instantiated in this context.

5.2 Formalization of Timed CSP

In this section, first we present our formalization of the syntax and the operational semantics of Timed CSP in the Isabelle/HOL theorem prover. Then, we show how the previously presented verification techniques can be inherited for Timed CSP by interpreting Timed CSP as a timed LTS. Furthermore, we show that all considered kinds of bisimulation are observational congruences in the context of Timed CSP. Finally, we give an outlook towards the integration of Timed CSP's denotational semantics into our formalization by operationally characterizing the timed traces of processes and by verifying their relationship to weak timed bisimulation.

5.2.1 Syntax

We realize the syntax of Timed CSP processes by defining an inductive datatype ('v,'a)Process, which is parameterized over the type variable 'v representing process variables and the type variable 'a representing the process alphabet. An excerpt of the datatype definition is given in the following figure.

```
datatype ('v,'a)Process =
    STOP | SKIP | Variable "'v"                    (<_>)
  | Prefix "'a" "('v,'a)"                          (infixr →)
  | Menuchoice "'a set" "'a ⇒ ('v,'a)Process"      (?: _ → _)
    ...
```

The datatype constructors correspond to Timed CSP operators. However, to give Timed CSP operators their usual look, we additionally introduce syntax abbreviations. This means, for example, that process variables within a Timed CSP process are denoted <v> instead of Variable v or that the *Prefix* operator can be written as a → P instead of Prefix a P.

Formalization of Timed CSP in the Isabelle/HOL Theorem Prover

We introduce common derived process operators such as *WAIT*, *Delayed Prefix* and the *Guard* process separately by using the concept of syntactical abbreviations in Isabelle.

```
abbreviation WAIT :: real ⇒ ('v,'a) Process
where WAIT d ≡ STOP ▷^d SKIP

abbreviation delayedPrefix :: 'a ⇒ real ⇒ ('v,'a) Process
                           ⇒ ('v,'a) Process        (infix →-)
where a →^d P ≡ a → (STOP ▷^d P)

abbreviation guard :: bool ⇒ ('v,'a) Process
                    ⇒ ('v,'a) Process               (_ && _)
where g && P ≡ (if g then P else STOP)
```

Process variables are given an interpretation by a process variable assignment of the type 'v ⇒ ('v,'a)Process mapping process variables to processes. Furthermore, we interpret Timed CSP processes with respect to a fixed set of observable events. The intuition is that, although hidden, observable events can still be distinguished in the timed LTS given by the operational semantics. We explain this mechanism in the next subsection in more detail. We define a locale for Timed CSP where an abstract process variable assignment and an abstract set of observable events is fixed.

```
locale tcsp
  fixes asg :: "'v ⇒ ('v,'a) Process"
    and obs :: "'a set"
```

Process variables in combination with respective process variable assignments are used to conveniently describe (mutually) recursive processes. Consider, for example, the syntactical definition of a simple timed counter process. It is a variant of the counter process described in Section 2.2.1. Using our formalization, we can syntactically formalize it as follows.

5.2 Formalization of Timed CSP

Counter Process Example First, we need to introduce a datatype definition for the process variables of the counter process and a datatype definition for the process alphabet.

```
datatype vars = Count "nat"

datatype ev = succ | pred
```

This datatype implicitly introduces infinitely many process variables. The state of the counter is parameterized with an (arbitrary) natural number, which is used to count the number of communicated succ events for which no pred event has been communicated yet.

Now we can assign a process definition to each process variable by defining the following process variable assignment.

```
fun  pasg :: vars ⇒ (vars,ev)Process
where
  pasg (Count 0) = succ →¹ <Count 1>
| pasg (Count n) = succ →¹ <Count (n + 1)>
                 □ pred →¹ <Count (n − 1)>
```

To interpret this process definition within Timed CSP it is then necessary to interpret the tcsp locale appropriately.

```
interpretation counter: tcsp "pasg" "{}"
```

Note that we take the set of observable events as the empty set. As no *Hiding* is involved in the process definitions, the choice of this set is actually of no meaning here.

Until now, we have presented the possibilities to formalize Timed CSP processes syntactically using our formalization of Timed CSP in the Isabelle/HOL theorem prover. To give processes a formal semantics, we consider the operational semantics. In the following, we describe how we formalized the operational semantics of Timed CSP in the Isabelle/HOL theorem prover.

Formalization of Timed CSP in the Isabelle/HOL Theorem Prover

5.2.2 Operational Semantics

Within the context of the `tcsp` locale, we are now able to define the operational semantics of Timed CSP as explained below. However, first we present a slight extension of the semantics of Timed CSP, which makes hidden events visible to a certain degree. This has the advantage that we can introduce arbitrary observation points on a given Timed CSP design. This eases verification and makes internal behavior more transparent to the designer. At the same time, it enables the designer to hide visible events, which makes them urgent and thus enforces timing behavior, without losing observability.

Making Timed CSP Urgent In the original version of Timed CSP, only internal τ events have priority over timed steps. This means that visible events can be delayed until the next internal event gets possible. The idea is that a Timed CSP process models the timed behavior of processes that are not yet connected to some environment. An event only occurs when the environment is willing to communicate the visible event. To close the system, the *Hiding* operator of Timed CSP is used, meaning that the events in the hidden set are not influenced by a possible environment any more. Consequently, hidden events are not visible to some environment and therefore transformed to τ events. The problem with this treatment is that the internal behavior of a process cannot be analyzed because the only information propagated by the operational semantics is τ.

We slightly extended our formalization to enable the analysis of the internal behavior of a process by globally introducing a set *obs* of observable events, which remain visible even after being hidden. To achieve this, we define the following variants of the original rule concerning the *Hiding* operator.

5.2 Formalization of Timed CSP

$$\frac{P \xrightarrow{a} Q}{P \setminus A \xrightarrow{\tau_a} Q \setminus A} \quad a \in A \wedge a \in obs$$

$$\frac{P \xrightarrow{a} Q}{P \setminus A \xrightarrow{\tau} Q \setminus A} \quad a \in A \wedge a \notin obs$$

This is a conservative extension of the original operational semantics in the sense that the timed LTS is the same as before except that transitions labeled with τ are possibly replaced by some τ_a. In the case that $obs = \emptyset$, both transition systems are the same. We have chosen this type of conservative extension in order to keep the convenient properties of the original operational semantics.

To define the operational semantics of Timed CSP, we consider a common type for labels of the underlying transition system. There are five different kinds of transition steps: event steps, internal unobservable steps, internal observable steps, terminating steps and timed steps. We encode the corresponding labels using the datatype `'a eventplus`, where `'a` denotes the process alphabet.

```
datatype 'a eventplus = ev "'a"
                     | tau              (τ)
                     | urgent "'a"      (τ_)
                     | tick             (√)
                     | time "real"
```

We define the semantics using two inductive sets of triples, which share the same type `(('v,'a)Process , 'a eventplus) lts`. Event steps define the transition steps with respect to the labels `ev a`, τ, τ_a, and $\sqrt{}$. Timed steps only define the steps with respect to the label `time t`. An excerpt of the semantics definition is given in Figure 5.1 and Figure 5.2.

Formalization of Timed CSP in the Isabelle/HOL Theorem Prover

```
inductive_set  evstep ::  (('v,'a)Process , 'a eventplus)lts
  ...
| Seqcompo_step1: ⟦ (P , e , P') ∈ evstep ; e=ev a ∨ internal e ⟧
                  ⟹ ((P;;Q) , e , (P';;Q)) ∈ evstep
| Seqcompo_step2: ⟦ (P , √ , P') ∈ evstep ⟧
                  ⟹ ((P;;Q) , τ , Q) ∈ evstep
  ...
| Hiding_step1: ⟦ (P , ev a , P') ∈ evstep ; a∈A ; a∈obs ⟧
                ⟹ (P\A , τ_a , P'\A) ∈ evstep
| Hiding_step2: ⟦ (P , ev a , P') ∈ evstep ; a∈A ; a∉obs ⟧
                ⟹ (P\A , τ , P'\A) ∈ evstep
| Hiding_step3: ⟦ (P , ev a , P') ∈ evstep ; a∉A ⟧
                ⟹ (P\A , ev a , P'\A) ∈ evstep
| Hiding_step4: ⟦ (P , e , P') ∈ evstep ; e=√ ∨ internal e ⟧
                ⟹ (P\A , e , P'\A) ∈ evstep
  ...
| Timeout_step1: ⟦ (P , e , P') ∈ evstep ; e=ev a ∨ e=√ ⟧
                 ⟹ ((P ▷$^d$ Q) , e , P') ∈ evstep
| Timeout_step2: ⟦ (P , e , P') ∈ evstep ; internal e ⟧
                 ⟹ ((P ▷$^d$ Q) , e , (P' ▷$^d$ Q)) ∈ evstep
| Timeout_step3: ((P ▷$^0$ Q) , τ , Q) ∈ evstep
  ...
| Variable_step: (<v> , τ , asg a) ∈ evstep
```

Figure 5.1: Excerpt of the Event Step Definition in Isabelle/HOL

```
inductive_set  tstep ::  (('v,'a)Process , 'a eventplus)lts
  ...
| Seqcompo_step: ⟦ (P , time t , P') ∈ tstep ;
                   ¬(∃ R. (P , √ , R) ∈ evstep) ⟧
                 ⟹ (P;;Q , time t , P';;Q) ∈ tstep
  ...
| Hiding_step: ⟦ (P , time t , P') ∈ tstep ;
                 ∀ a∈A. ¬(∃ Q. (P , ev a , Q) ∈ evstep) ⟧
               ⟹ (P\A , time t , P'\A) ∈ tstep
  ...
| Timeout_step: ⟦ (P , time t , P') ∈ tstep ; t≤d ⟧
                ⟹ ((P ▷$^d$ Q) , time t , (P' ▷$^{d-t}$ Q)) ∈ tstep
  ...
```

Figure 5.2: Excerpt of the Timed Step Definition in Isabelle/HOL

5.2 Formalization of Timed CSP

Note that the operational rules for timed transitions depend on the (previously defined) event transitions. Internal events are instantaneous in Timed CSP. This means that time may not advance if internal transitions are enabled. In the semantics of *Sequential Composition* and *Hiding*, it is thus necessary to allow time to advance only if internal transitions are not enabled (see [Sch99]). This is the case, for example, if the first process of the *Sequential Composition* can successfully terminate.

Compared to the original semantics of Timed CSP as given in Section 2.2.2, the operational semantics here differs by the special treatment of urgent τ_a steps. However, as for example in rule `Timeout_step2`, they are treated like τ in the original semantics, i.e., they do not, for example, resolve the *Timeout* operator. To achieve this, we have introduced the predicate `internal`, which is true for τ and τ_a labels.

The full operational semantics of Timed CSP is given by the union of event steps and timed steps. We further provide the syntactical abbreviation P -<e>→ P' to denote (P , e , P') ∈ step.

```
definition step:: (('v,'a)Process,'a eventplus)lts (infix "-<_>→")
where  step ≡ evstep ∪ tstep
```

In order to increase the automatization in proofs, we derive a compound lemma for deducing steps of Timed CSP processes.

```
lemmas step_simps = STOP_simps,SKIP_simps,...,Hiding_simps,...
```

Each of the single lemmas expresses which processes can be reached by a single step beginning from a particular Timed CSP operator. For example, `Hiding_simps` expresses the following.

Formalization of Timed CSP in the Isabelle/HOL Theorem Prover

```
lemma Hiding_simps: "P\A —<e>→ Q ⟷
    (∃a. e=τ_a ∧ a∈A ∧ a∈obs ∧ (∃P'. Q=P'\A ∧ P —<ev a>→ P'))
  ∨ (∃a. e=τ   ∧ a∈A ∧ a∉obs ∧ (∃P'. Q=P'\A ∧ P <ev a>→ P'))
  ∨ (∃a. e=ev a ∧ a∉A ∧ (∃P'. Q=P'\A ∧ P —<ev a>→ P'))
  ∨ ((e=√ ∨ internal e) ∧ (∃P'. Q=P'\A ∧ P —<e>→ P'))
  ∨ (∃t. e=time t ∧ (∀ a∈A. ¬(∃ Q. P —<ev a>→ Q)) ∧
              (∃P'. Q=P'\A ∧ P —<time t>→ P'))"
```

This compound lemma is very useful to reason about reachable states of a process. If the compound lemma `step_simps` is given to the Isabelle simplifier tactic `simp`, the reachable steps of a composed process can be deduced in an automatized fashion because the simplifier continuously applies matching sublemmas of the `step_simps` compound lemma. The result is a disjunction corresponding to possible steps of the process. Considering each disjunct separately enables reasoning about the reachable states one after another.

Properties of the Operational Semantics

Based on this formalization of the operational semantics of Timed CSP, we can verify important properties [Sch99] with respect to the operational semantics of Timed CSP as defined by `step`. This, on the one hand, shows that the operational semantics is correctly formalized and, on the other hand, gives useful properties that are needed for later proofs.

Urgency of Internal Events: It is important that internal steps have priority over timed steps. This property is captured by a lemma that states that timed steps are disabled if internal steps are possible. Note that external events cannot be forced to occur at a certain point in time. They need to be internalized by the *Hiding* operator to achieve urgency.

5.2 Formalization of Timed CSP

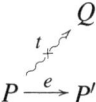

```
lemma internal_urgency: [ P –<e>→ P' ; internal e ]
    ⟹ ¬(∃ Q t. P –<time t>→ Q)
```

Constancy of Offers: Timed steps do not change the set of offered (visible) events of a process, i.e., only events (internal as well as external) may change the offers of a process.

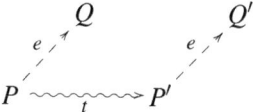

```
lemma constancy_offers: [ P –<time t>→ P' ; e=√ ∨ (∃a. e=ev a) ]
    ⟹ (∃ Q. P –<e>→ Q) ⟷ (∃ Q. P' –<e>→ Q)
```

Time Determinism: Timed steps do not introduce further nondeterminism. This means that every two timed steps of the same duration will reach the same target process.

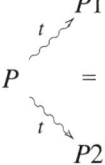

```
lemma time_determinism: [ P –<time t>→ P1 ; P –<time t>→ P2 ]
    ⟹ P1=P2
```

131

Formalization of Timed CSP in the Isabelle/HOL Theorem Prover

Time Additivity: Two consecutive timed steps may be summarized into one timed step (of the summed duration). The other direction holds as well. Every timed step may be divided into two consecutive timed steps.

$$P1 \xrightarrow{t1+t2} P3 \qquad P1 \xrightarrow{t1} P2 \xrightarrow{t2} P3$$

```
lemma time_add: (∃ P2. P1 -<time t1>→ P2 ∧ P2 -<time t2>→ P3)
              ⟷ (P1 -<time (t1+t2)>→ P3 ∧ t1 >0 ∧ t2 >0)
```

All proofs above have been performed using induction over the Timed CSP processes and using the definition of the operational semantics.

5.2.3 Timed CSP as a Timed Labeled Transition System

The operational semantics of Timed CSP defines a timed LTS. This can formally be shown by verifying that the locale `tcsp` is a sublocale of `timed_lts`.

```
sublocale tcsp ⊆ timed_lts step τ time
```

This statement introduces the proof obligation that the operational semantics of Timed CSP provides the assumptions of the timed_lts locale, which we have verified using the properties from the previous subsection.

After successfully finishing the sublocale proof, all definitions and lemmas that are provided within the context of timed LTSs are also available in the context of Timed CSP. This includes especially the considered kinds of bisimulations, our timed HML, and their interrelationship.

5.2 Formalization of Timed CSP

5.2.4 Bisimulation as an Observational Congruence

We have verified that all three considered kinds of bisimulation are observational congruences with respect to Timed CSP processes in the Isabelle/HOL theorem prover. This means that it is possible to conclude the bisimilarity of two composed processes $P \oplus P'$ and $Q \oplus Q'$ (for some Timed CSP operator $\oplus$) from the bisimilarity of their subprocesses, i.e., the bisimilarity of P and Q and the bisimilarity of P' and Q'. We performed the proof for this in Isabelle/HOL by induction over Timed CSP processes and by using additional lemmas concerning possible complex transition steps of composed processes.

The crucial part of the proof was to verify that *Sequential Composition* and *Hiding* also fulfill the congruence property. The main difficulty stems from the fact that under weak (timed) bisimilarity, a simple timed step may be answered by a complex step, which may consist of arbitrarily many internal transitions and possibly also of arbitrarily many timed steps. For the proof, the complex step must be decomposed and it has to be shown that after each of the arbitrarily many steps, it is not possible to communicate the termination event $\sqrt{}$ (for *Sequential Composition*) or hidden events (for *Hiding*), respectively.

As an example, one part of the congruence proof for weak bisimulation with respect to *Sequential Composition* is as follows. Assume that P and Q are weak bisimilar and we want to show that $P;R$ is weak bisimilar to $Q;R$ for some arbitrary process R. A timed step of $P;R$ can only stem from a timed step of P where P cannot terminate. If P evolves to P' by some timed step with duration t, it follows from the bisimilarity of P and Q that also Q can evolve to some Q' in t time units (where also internal steps are allowed) such that P' and Q' are bisimilar. In summary, we have

- P and Q are weak bisimilar

Formalization of Timed CSP in the Isabelle/HOL Theorem Prover

- $P \xrightarrow{t} P'$
- $\neg(P \xrightarrow{\checkmark})$
- $Q \xrightarrow{\tau} \ldots \xrightarrow{\tau} Q_1 \xrightarrow{t} Q_2 \xrightarrow{\tau} \ldots \xrightarrow{\tau} Q'$
- such that P' and Q' are weak bisimilar

Now we would like to show that the composed process $Q;R$ can evolve to $Q';R$ in t time units. Formally, we have to show

$$Q;R \xrightarrow{\tau} \ldots \xrightarrow{\tau} Q_1;R \xrightarrow{t} Q_2;R \xrightarrow{\tau} \ldots \xrightarrow{\tau} Q';R$$

The challenge is that the timed step (from $Q_1;R$) in the evolution of $Q;R$ is only allowed if no terminating step ($\checkmark$) of Q_1 is possible. The proof for this takes into account that P cannot carry out an internal step, this only being possible if a timed step is not possible. Thus, due to bisimilarity of P and Q, every internal step of Q must be answered by an empty τ-sequence of P leading trivially again to P. By induction, it follows that every process that is reachable from Q by only doing τ steps is weak bisimilar to P. This implies that process Q_1 (where the timed step is done in the evolution of Q) must be weak bisimilar to P. Finally, we can conclude that Q_1 cannot perform a terminating $\checkmark$-step because P cannot terminate. The overall proof relies on the fact of *internal urgency*. A similar proof is needed in the context of weak timed bisimulation. There, the properties *time additivity* and *constancy of offers* of Timed CSP's operational semantics, as explained in Section 5.2.2, are additionally used.

5.2.5 Outlook: Denotational Semantics

We have presented our formalization of the operational semantics of Timed CSP in Isabelle/HOL. In future work, it would be very convenient to consider the denotational semantics of Timed CSP as well in order to be able

5.2 Formalization of Timed CSP

to establish proofs on a more abstract semantic domain. Our formalization gives a useful starting point for this task. The denotational semantics can be operationally characterized using the operational semantics as shown, for example, in [Sch95]. In the following, we present our formalization of the timed traces semantics in the context of timed LTSs and show their relationship to bisimulations. Furthermore, as an example, we verify that terminating steps can only occur at the end of a timed trace in the context of Timed CSP.

Timed Traces Semantics

A timed trace is a sequence of timed events where the time points are relative to the beginning of the considered trace. To characterize timed traces operationally, we first define timed behaviors of processes as a sequence of compound time-event steps of the underlying timed LTS.

```
primrec isBeh :: 's ⇒ (real×'a)list" ⇒ bool
where
  "isBeh P [] = True"
| "isBeh P ((t,a)#xs) = ∃ P'. P ↝^(t,a) P' ∧ isBeh P' xs
```

To define the timed traces of a process, we need to transform its timed behaviors such that we get the time stamps of events relative to the beginning of each timed behavior.

```
primrec transform_h :: real ⇒ (real×'a)list ⇒ (real×'a)list
where
  transform_h r [] = []
| transform_h r ((t,a)#xs) = (t+r,a) # transform_h (t+r) xs

definition transform :: (real×'a)list ⇒ (real×'a)list
where  transform xs ≡ transform_h 0 xs
```

Finally, the set of timed traces is defined by the transformation of all possible behaviors of a process.

Formalization of Timed CSP in the Isabelle/HOL Theorem Prover

```
definition ttraces :: 's ⇒ (real×'a) list set
where ttraces P ≡ transform ' {b. isBeh P b}
```

Based on the definition of timed traces, we have been able to verify standard properties of them, like the prefix closure of timed traces, easily.

```
lemma ttraces_prefix_closure: "⟦ tr ∈ ttraces P ; tr'≤tr⟧
                              ⟹ tr'∈ttraces P"
```

Whenever `tr` is a timed trace of some process P and `tr'` is a prefix of `tr` ($tr' \leq tr$), then `tr'` is also a timed trace of P.

Another common property is that within the context of Timed CSP a terminating event can only occur at the end of a timed trace.

```
context tcsp
begin
  ...
lemma tick_final: "⟦ tr ∈ ttraces P ; √ ∈ event ' (set tr) ⟧
                  ⟹ √ = event (last tr)"
```

In this context, the function `event` selects the event from a time-event pair, i.e., `event(t,a) = a`.

An important property of timed traces is their relationship to weak timed bisimulations: Weak timed bisimilar processes have the same set of timed traces.

```
lemma weakt_bisim_ttrace_eq: "(P,Q) ∈ weak_timed.bisimilar
                             ⟹ ttraces P = ttraces Q"
```

This, for example, allows us to use the FDR2 refinement checker within our framework. In [Oua01], a discretization approach has been developed with which, for example, timed traces refinement can be shown using the FDR2 refinement checker. If the system under verification is finite, this can be done fully automatically. When applying our framework to parameterized real-time systems, we use FDR2 to verify timed traces equivalence of instances of the abstract and concrete parameterized models. Within our Isabelle/HOL formalization, we aim at showing weak timed bisimilarity between the abstract and concrete systems for arbitrary network sizes. So, a previous successful check that instances are timed traces equivalent gives evidence that the bisimilarity proof is principally possible. If the timed traces checks were not successful, the final bisimilarity proof could not be established, as bisimilarity implies timed traces equivalence for all instances.

So far, we have presented the formalization of the verification techniques of bisimulations and our timed HML. Furthermore, we have shown how we formalized the operational semantics of Timed CSP and how the previous verification techniques are inherited for Timed CSP. Additionally, we formalized the timed traces of processes and verified some common properties of them. In the next section, we outline some important mechanisms that we have included in our formalization in order to describe and verify parameterized real-time systems.

5.3 Coping with Parameterized Systems

The main goal of applying our presented framework is to mechanically prove (possibly infinite) parameterized real-time systems correct with respect to given requirements. Therefore, on the one hand, we need to describe and reason about arbitrarily large networks of processes and, on the other hand, we need to express properties of parameterized systems in our timed HML. In the following, we briefly explain some important mechanisms in our formalization to achieve this goal.

5.3.1 Arbitrarily Large Networks of Timed Processes

To describe arbitrarily large networks of timed processes, we formalized a variant of the indexed alphabetized parallel composition[1] of processes and provide lemmas to reason about these systems. So, we have included the operator $P|[A,B]|Q$ in our syntax, which has the semantics that P and Q behave independently but must synchronize on shared events $\mu \in A \cap B$ and on $\sqrt{}$. Furthermore, P and Q are only allowed to communicate events from A or B, respectively.

We define a version of indexed alphabetized parallel composition using the following definition.

```
fun compo::  "nat ⇒ (nat⇒'a set) ⇒ (nat⇒('v,'a)Process)
                 ⇒ ('v,'a)Process"
where
  compo 1       alph P = P 1
| compo (n+1)   alph P = let A = ⋃{x. ∃i. i≥1 ∧ i≤n. x=alph i} ;
                             B = alph (n+1)
                         in (compo n alph P) |[A,B]| P(n+1)
```

Note that we assume that the network has at least size one. We have introduced the convenient syntax `|| k* [alph] P` for this definition. Its meaning is that all processes P(1) , ..., P(k) behave independently but must synchronize on their shared events with respect to the given process alphabet `alph`. Formally, this is verified in the following lemma, which allows for deducing possible steps of arbitrarily large networks.

---
[1] In textbooks of (Timed) CSP, this operator is introduced using a finite indexing set I: $\|_{A_i}^{i \in I} P_i$

5.3 Coping with Parameterized Systems

```
lemma step_compo: ⟦ || k∗ [alph] P −<e>→ Q ; k>0 ⟧ ⟹
(∃a. e=ev a ∧
    (∃ P' I. I ≠ {} ∧ (∀ i∈I . i≥1 ∧ i≤k )
       ∧ (∀ i . i≥1 ∧ i≤k ⟶ (i∈I ⟶ P i −<ev a>→ P' i)
                              ∧ (i∉I ⟶ P' i = P i))
       ∧ (k≥2 ⟶ (∀i. i≥1 ∧ i≤k ⟶ (i∈I ⟶ a∈alph i)
                                    ∧ (i∉I ⟶ a∉alph i )))
       ∧ Q = || k∗ [alph] P'))
∨ (internal e ∧
    (∃ i P1. i≥1 ∧ i≤k ∧ P i −<e>→ P1
       ∧ Q = || k∗ [alph] (λj. if j=i then P1 else P j) ))
∨ (e=√ ∧
    (∃ P'. (∀ i. i≥1 ∧ i≤k ⟶ P i −<√>→ P' i)
       ∧ Q = || k∗ [alph] P'))
∨ (∃ t. e=time t ∧
    (∃ P'. (∀ i. i≥1 ∧ i≤k ⟶ P i −<time t>→ P' i)
       ∧ Q = || k∗ [alph] P'))
```

If an event ev a is communicated then there exists a set I of process indices such that the corresponding processes engage in the event ev a and the other processes stay as before. Furthermore, all processes referenced by I have event a in their alphabet and the others have not. An internal step is only performed by a single process of the network. Finally, terminating steps and timed steps need to be performed by all processes in the network synchronously.

5.3.2 Expressing Properties of Parameterized Real-Time Systems

To conveniently express crucial properties of parameterized real-time systems, we define a finite universal quantifier for our timed HML by the following definition.

```
function finAll :: nat ⇒ nat ⇒ (nat⇒'b formula)
                 ⇒ 'b formula (⋀[_,_] _)
where
   ⋀[i,j] ϕ = (if i>j then tt
                      else ϕ i ∧f (⋀[i+1,j] ϕ))
```

Under the assumption that there exist events corresponding to single processes of network processes, this operator allows for adjusting logical properties depending on the index of the process.

Let, for example, c be a channel with process indices as possible values and n be the currently considered network size. Then, we can, for example, express properties like (1) if c.i is currently possible then formula $\phi(i)$ has to hold or (2) if some c.i is possible then it has to occur after at least $d_1(i)$ time units.

1. ⋀[1,n] λ i. ⟨⟨ c.i ⟩⟩$_{=0}$ tt $\longrightarrow_f$ $\phi(i)$

2. ⋀[1,n] λ i. [[c.i]]$_{\geq d_1(i)}$

Note that the finite universal quantifier does not require an extension of our timed HML datatype. Thereby, we keep a very small but expressive timed logic. In Chapter 7, we apply this logic in the context of our real-time scheduler case study and thereby show that it is expressive enough to describe crucial properties of parameterized real-time systems.

5.4 Summary

In this chapter, we have presented our Isabelle formalization of Timed CSP and the generic Isabelle formalizations of bisimulations and our timed Hennessy-Milner logic enhanced with coinductive invariants, which are both based on (timed) labeled transition systems. We have verified that the invariant satisfaction of timed HML formulae is preserved by weak timed

5.4 Summary

bisimulation, which gives a powerful proof principle for our framework for the verification of parameterized real-time systems. By interpreting the operational semantics of Timed CSP as a timed labeled transition system, we inherit these verification techniques and related lemmas for Timed CSP. The advantage is that our framework is thereby also able to cope with other process algebras as long as they can also be interpreted as timed labeled transition systems. All considered kinds of bisimulations in the context of Timed CSP have the convenient property to be observational congruences, which is very useful for compositional verification. In Chapter 7, we show how these properties help to verify our scheduler case study. To facilitate the verification using the FDR2 refinement checker, we have presented our formalization of timed traces and their relationship to weak timed bisimulation. In future work, it would be interesting to also characterize compositional denotational semantics of Timed CSP such as timed failures semantics as discussed in Chapter 8. Our timed traces semantics can serve as a good starting point for this. Finally, we have presented helpful mechanisms and lemmas to conveniently describe and reason about parameterized real-time systems.

Our formalization of Timed CSP and the presented proof techniques in the Isabelle/HOL theorem prover enable the mechanical verification of Timed CSP processes. This has the advantage that corner cases cannot be overlooked in proofs. Moreover, our formalization copes with infinite systems like parameterized systems. Therefore, not only correct proofs are ensured but proofs can also be performed about infinite systems, which automatic verification tools like, for example, a model checker cannot perform. The effort of mechanical verification as performed in an interactive theorem prover is of course higher than for paper-and-pencil proofs or the verification in a model checker. However, there is a need for absolute correct proofs about (parameterized) real-time systems especially in the area of safety-critical embedded systems that only an interactive proof environ-

Formalization of Timed CSP in the Isabelle/HOL Theorem Prover

ment can give. To reduce the effort of interactive theorem proving to a certain degree, instances should be first automatically analyzed such that potential errors are detected early. To achieve this, we provide transformations from Timed CSP to timed automata and to tock CSP as described in the following chapter.

6 Integration of Automatic Verification Tools

Within our verification framework, we aim at automatically verifying instances of parameterized real-time systems before entering the relatively time-consuming phase of comprehensive verification. Currently, there exists only little tool support for the automatic verification of Timed CSP processes. However, subsets of Timed CSP can be transformed to other formalisms that support automatic verification. We consider two of these approaches in this thesis, which we need, however, to adapt in order to cope with the (original) semantics of Timed CSP correctly. Furthermore, we provide extensions of these transformations to allow a larger class of Timed CSP processes to be transformed.

The first approach defines a transformation from Timed CSP processes to timed automata. This approach is presented in [DHQ$^+$08]. However, the transformation rules concerning *External Choice* and *Timeout* given there contain flaws. We therefore give corrected transformation rules for these process operators. Furthermore, we provide transformation rules

Integration of Automatic Verification Tools

for *Hiding* and the general *Interrupt* operator, which were not considered in [DHQ+08]. In [Wu10], we have implemented our transformation rules to enable the automatic transformation of Timed CSP processes to timed automata. We target the dialect of UPPAAL timed automata such that the UPPAAL model checker [BY04] can be used for verification purposes and the UPPAAL simulator can be used for simulation and debugging purposes.

In the second approach, Timed CSP processes are mapped to a discretely timed dialect of CSP, called tock CSP. The "trick" is to discretize the time domain of Timed CSP. The progress of time is modeled by performing discrete *tock* events and not by performing continuous timed steps any more. This approach was presented in [Oua01], where a syntactic mapping of Timed CSP processes to tock CSP processes is presented. However, the underlying semantical model of (Timed) CSP considered there differs from the original semantics with respect to termination of parallel processes. In [Oua01], a parallel process can successfully terminate if *one* of its subprocesses can. In the original semantics of (Timed) CSP, parallel processes can only terminate if *all* its subprocesses can. Therefore, we have adapted the transformation rules with respect to *External Choice*, where this different treatment poses problems. Furthermore, we provide transformations for *Interrupt* and *Timed Interrupt*, which were not considered in [Oua01]. We have implemented the transformation rules in [Zho10] to enable the automatic transformation of Timed CSP processes to tock CSP. As a result, we are able to employ the FDR2 refinement checker [GRA05], which was originally developed for the verification of untimed CSP. However, it also supports the verification of tock CSP processes with respect to traces refinement in its tau-priority model.

Note that both our tools can handle finite Timed CSP processes only. By this, we mean that the semantical state space of a Timed CSP process is finite if we leave out intermediate states introduced by timers within the

Timeout, Timed Interrupt, and *WAIT* processes. For example, the process $R = a \rightarrow STOP \stackrel{5}{\triangleright} b \rightarrow STOP$ is finite because the semantical state space is only infinite due to intermediate states that stem from the timer in the *Timeout* process. However, the recursive process $R = a \rightarrow^d R \,|||\, b \rightarrow STOP$ is an infinite process because after each occurrence of event a, the term $|||\, b \rightarrow STOP$ is added to the currently evolving process in the semantical state space.

We have integrated our transformation engines into our framework for the mechanical verification of parameterized real-time systems. They are used to transform finite instances of parameterized real-time systems to timed automata and tock CSP. With that we can verify properties using the UPPAAL model checker and check the semantical equivalence of instances of the abstract and the concrete parameterized real-time systems using the FDR2 refinement checker.

The advantage of our transformation engines from Timed CSP to timed automata and to tock CSP is that the existing established verification tools UPPAAL and FDR2 can be used to verify Timed CSP processes automatically. Thus, both orthogonal kinds of properties can be verified on instances of a parameterized real-time system. By this, we get a good evidence that also the respective proofs in the comprehensive verification phase are possible. In the case that these properties are not valid, counterexamples are provided, which can be used for debugging.

In this chapter, we present our adapted and extended transformations from Timed CSP to timed automata and from Timed CSP to tock CSP, which we prototypically implemented. In Section 6.1, we review the transformation from Timed CSP to timed automata as given in [DHQ$^+$08] and describe our adaptations and extensions. Furthermore, we sketch how properties given in our timed Hennessy-Milner logic (HML) can be verified on the resulting model using the UPPAAL model checker. In Section 6.2, we review the transformation from Timed CSP to tock CSP, which is presented in [Oua01], and describe our adaptations and extensions. Then, we discuss how semantical equivalence with respect to timed traces can be shown on the resulting models using the FDR2 refinement checker. This chapter closes with a summary concerning the transformations of Timed CSP to timed automata and tock CSP and their integration into our verification framework in Section 6.3.

6.1 Transformation from Timed CSP to Timed Automata

In [DHQ$^+$08], a transformation from Timed CSP to timed automata is presented, which largely fits our needs. However, as described above, we had to adapt and extend the corresponding transformation rules. Still, some assumptions are necessary in order to be able to transform Timed CSP processes correctly into equivalent timed automata. In Section 6.1.1, we introduce necessary assumptions on Timed CSP processes for the transformations to be correct. We present the transformation rules with our adaptations and extensions in Section 6.1.2. In Section 6.1.3, we discuss how timed HML properties can be verified on a translated Timed CSP process using the UPPAAL model checker. Finally, in Section 6.1.4, we summarize the presented transformation approach.

We follow [DHQ$^+$08] by slightly extending the syntax of timed automata as given in Section 2.3 by additionally considering final locations. These have no relevance for the semantics of timed automata but are used as anchors to define especially the transformation rule for *Sequential Composition*.

6.1 Transformation from Timed CSP to Timed Automata

Definition 18 (Intermediate Representation of Timed Automata) *The syntax of an intermediate timed automaton final locations and is given by the tuple $(L, l_0, F, C, \Sigma, E, I)$ where*

- *L is a finite set of locations and $l_0 \in L$ is the initial location.*
- *F is a set of final locations.*
- *C is a finite set of clock variables.*
- *Σ is a finite set of actions with $\tau \in \Sigma$ denoting an internal transition.*
- *$E \subseteq L \times B(C) \times \Sigma \times \mathbb{P}(C) \times L$ is a finite set of edges. We also use the notation $l \xrightarrow{(g,a,r)} l'$ for $(l, g, a, r, l') \in E$.*
- *$I :: L \to B(C)$ is a function assigning an invariant to each location.*

Note that in the following presentation of transformation rules, we use a slightly extended definition of intermediate timed automata, where locations can be marked as urgent and integer variables can be introduced that are possibly manipulated on edges and that can be used in guards of edges. Their meaning is the same as in UPPAAL timed automata.

Our overall transformation algorithm works as follows. If a process $P1 \oplus P2$ is to be translated where $\oplus$ is some Timed CSP operator, then first translate $P1$ and $P2$ such that the resulting automata representing these subprocesses are A_{P1} and A_{P2}. Then, combine A_{P1} and A_{P2} according to the corresponding transformation rule. The transformation rules for all process operators are described below. We mostly informally sketch the transformation rules, for their formal treatment we refer to [DHQ[+]08].

6.1.1 Assumptions

To get correct results, we need to make the following assumptions.

Integration of Automatic Verification Tools

- The presented transformation assumes that all timed operators in the Timed CSP process to be transformed are defined over natural numbers. This means, for example, that for a *WAIT(n)* process being part of the overall process to be translated, it is assumed that $n \in \mathbb{N}$.

- The treatment of termination by using final locations makes it necessary that each *SKIP* in the process to be transformed occurs within the left process of some *Sequential Composition*. So whenever a subprocess is able to terminate it can evolve to the second process of a *Sequential Composition*.

- We make the assumption that a Timed CSP process to be transformed is finite (as explained in the beginning of this chapter).

6.1.2 Transformation Rules

Under these assumptions, a Timed CSP process can be transformed into timed automata according to the following rules.

Basic Timed CSP Processes The process *STOP* is represented by a single initial location. There, no event can be performed but time can arbitrarily advance.

◯

The *SKIP* process is represented by an initial location that is final at the same time. So, *SKIP* actually behaves like *STOP*. However, as described below, within *Sequential Compositions* final locations are used for resolving *Sequential Composition*.

Ⓕ

6.1 Transformation from Timed CSP to Timed Automata

The representation of the *Prefix* process $a \rightarrow P$ takes the translated automaton A_P representing P and introduces a new initial location $init'$. An edge labeled with a leads to the initial location of A_P.

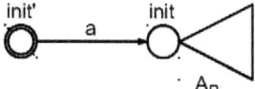

Internal Choice The *Internal Choice* construction $P1 \sqcap P2$ is represented by an urgent initial location with two τ-labeled edges leading non-deterministically to the initial location of either A_{P1} or A_{P2} representing $P1$ and $P2$, respectively.

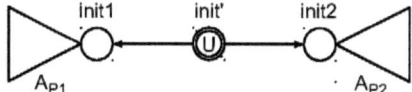

Parallel Composition We transform *Parallel Composition* by computing the syntactic cross product of the involved automata. In contrast to [DHQ+08], we consider synchronization sets explicitly. In the treatment there, synchronization is established over all shared events of the involved automata.

Let $A_i = (L_i, l_0^i, F_i, C_i, \Sigma, E_i, I_i)$ with $i \in \{1,2\}$ be two intermediate timed automata and S be an event set such that $S \subseteq \Sigma \setminus \{\tau\}$. Note that we assume for simplicity that both timed automata share the same alphabet Σ. Then we define

$$para(A_1, A_2, S) = (L_1 \times L_2, (l_0^1, l_0^2), F_1 \times F_2, C_1 \cup C_2, \Sigma, E, I)$$

with $I(l_1, l_2) = I_1(l_1) \wedge I_2(l_2)$, and with E defined by the smallest set satisfying

Integration of Automatic Verification Tools

- if $(l_1,g,e,r,l_1') \in E_1$ and $e \notin S$ then $((l_1,l_2),g,e,r,(l_1',l_2)) \in E$
- if $(l_2,g,e,r,l_2') \in E_2$ and $e \notin S$ then $((l_1,l_2),g,e,r,(l_1,l_2')) \in E$
- if $(l_1,g_1,e,r_1,l_1') \in E_1$ and $(l_2,g_2,e,r_2,l_2') \in E_2$ and $e \in S$ then $((l_1,l_2), g_1 \wedge g_2, e, r_1 \cup r_2, (l_1',l_2')) \in E$

Let $P1 \underset{S}{\parallel} P2$ be a Timed CSP process, then its representation in timed automata is given by $para(A_{P1}, A_{P2}, S)$.

External Choice In [DHQ$^+$08], the transformation of *External Choice* is realized by the syntactic cross product of A_{P1} and A_{P2} where the edges labeled with some visible event ev set an integer flag f to -1 if it stems from the first automaton or to 1 if it stems from the second automaton. Initially, f has the value 0. Furthermore, all edges corresponding to the first automaton get the additional guard $f \leq 0$ and edges corresponding to the second automaton get the additional guard $f \geq 0$. This means that both automata can perform internal events (or let time advance) as long as none of the involved automata has performed a visible event. When this happens, the edges of the other automaton are blocked by the respective guard. This treatment can introduce timelocks. For example, in the transformation of process $(WAIT(1) \Box a \rightarrow STOP); b \rightarrow STOP$, the timelock occurs if the second process performs its a event before 1 time unit passed by. Then, all edges of the first process are blocked by the guard $f \leq 0$, which is now unsatisfiable. Therefore, the urgent edge after having waited 1 time unit (introduced by the translation of *WAIT*) cannot be performed and the timelock occurs.

To correct this flaw and also for performance reasons, we change the transformation rule for *External choice* $(P1 \Box P2)$. In most cases, there are relatively few internal transitions until the choice is resolved in favor of the first process or the second process. Therefore, we first take the cross product of A_{P1} and A_{P2} but let visible events go to the respective target location

6.1 Transformation from Timed CSP to Timed Automata

in the automaton A_{P1} or A_{P2}. Then, we perform an analysis that detects unreachable locations in the crossproduct and deletes them. Thereby, only interleavings of initial internal steps are present in the parallel automaton. This heavily reduces the overall location set in many cases. Furthermore, after performing a visible event, which stems from the first or the second process, the composed automaton leaves the possibilities of the automaton corresponding to the other process. Thus, the flaw of [DHQ+08] does not occur in our transformation rule.

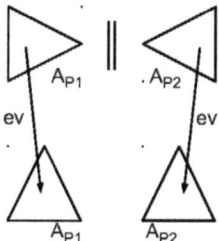

The cross product that is involved in the transformation rule is explained in the rule for *Parallel Composition*. The synchronization set in the transformation of *External Choice* is empty. The final locations of the composed automaton are the final locations of A_{P1} and A_{P2} and all cross product locations (l_1, l_2) where at least one of the locations l_1, l_2 is a final location.

Interrupt The *Interrupt* process $P1 \triangle P2$ is translated similarly to the *External Choice* process. We build a cross product of A_{P1} and A_{P2} and let event edges, which stem from $P2$, lead to the corresponding target location in A_{P2} thereby leaving the cross product. The final locations are given by the final locations of A_{P2} and all cross-product locations (l_1, l_2) where at least one of l_1, l_2 is a final location.

Integration of Automatic Verification Tools

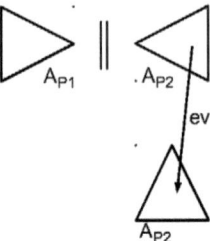

Sequential Composition The transformation of the *Sequential Composition* process $P1; P2$ is given by the two automata A_{P1} and A_{P2} representing $P1$ and $P2$ where the final locations of A_{P1} are connected to the initial location of A_{P2} and declared urgent. This means that the corresponding transition from A_{P1} to A_{P2} must happen before a next timed step, when the first process is able to terminate. The final locations of the composed automaton are given by the final locations of A_{P2}.

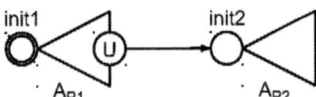

Timeout Consider the *Timeout* process $P1 \overset{t}{\triangleright} P2$. Then again, the transformed automata A_{P1} and A_{P2} representing the original processes $P1$ and $P2$ are taken to construct the composed automaton. An urgent initial location is introduced, which is connected to the initial location of A_{P1}. When taking this edge, a previously introduced fresh clock x is set to 0. Then, as long as the timeout t has not elapsed, the automaton A_{P1} can do internal steps and let some time advance. When a visible event is performed in A_{P1}, we update the flag *to_break* to *true* indicating that the *Timeout* is resolved in favor for the first automaton. Then, all edges leading to automaton A_{P2} are blocked and all invariants corresponding to the timeout t are irrelevant.

6.1 Transformation from Timed CSP to Timed Automata

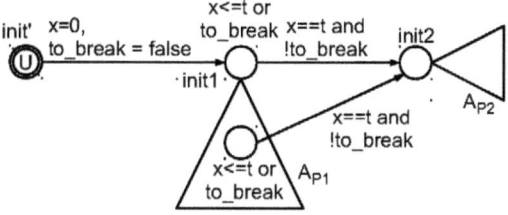

In the original transformation given in [DHQ+08], the behavior of the *Timeout* operator is not treated correctly. There, each step of the first process (also internal steps) disables the transitions to A_{P2}. Furthermore, if A_{P1} is recursive, it could reach its initial location again (even after performing a visible event). Then, the timeout t could elapse and A_{P2} could be reached. Another problem of the original treatment is that reaching again the initial location *init*1 could be blocked because in the meantime the timeout could have elapsed without having taken a transition to A_{P2}. With our transformation of *Timeout*, these problems are solved.

Timed Interrupt The treatment of *Timed Interrupt* is simpler than that of *Timeout*. For a Timed CSP process $P1 \triangle_t P2$, a fresh clock variable x that is initially set to 0 on the edge leading to the initial location of A_{P1}. An invariant $x \leq t$ is assigned to each location in A_{P1}. Furthermore, a τ-labeled edge with guard $x == t$ is added to each location of A_{P1}, which has the initial location of A_{P2} as its target. This ensures that the first automaton is left when the time specified in the *Timed Interrupt* elapses. When A_{P1} is able to terminate (before the time elapsed), it is in a final location. This situation must stem from a *SKIP* process, as this is the only process introducing final locations. Since we assume that each *SKIP* occurs in the left process of some *Sequential Composition*, the situation that A_{P1} is in a final location is handled by a transition to the second automaton of the *Sequential Composition*, thereby leaving A_{P1}.

Integration of Automatic Verification Tools

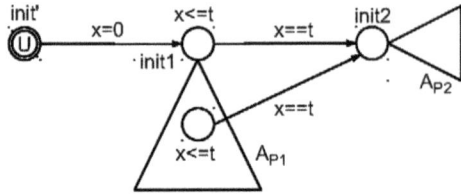

Hiding For the transformation of *Hiding* in a process $P \setminus A$, we set all source locations of edges labeled with some a out of A as urgent in A_P. Furthermore, we rename the hidden events using the predefined prefix _hidden_. This implies that events within the original Timed CSP process must not have this prefix in their names. The renaming of hidden events is necessary to ensure that hidden events do not, for example, resolve *External Choice* or *Timeout*. Another solution would be to delete synchronizations $a?$ for hidden events a and declare the corresponding source locations as urgent. This, however, has the disadvantage that hidden events cannot be considered in the later verification phase. Our transformation provides the correct handling of *Hiding* according to our extended operational semantics of Timed CSP where observable hidden events are urgent but not equal to τ.

Recursion In contrast to [DHQ$^+$08], we handle *Recursion* using process names instead of an explicit *Recursion* operator. To achieve this, we add a new labeling relation between locations and process names to our notion of an intermediate timed automaton. A process name appearing in a Timed CSP term is either translated with the automaton corresponding to the name, or, in case the name links to a process that is currently being translated, with a stub location. As soon as the translation of a named process is completed, the stub locations are linked back to the initial node of the corresponding automaton and are marked as resolved. At the end of the transformation there can be no remaining stub location. This way of tack-

6.1 Transformation from Timed CSP to Timed Automata

ling *Recursion* enables more sophisticated recursive definitions, on the one hand, and the detection of illegal forms of *Recursion*, on the other hand. For example, the process $P = pop \to STOP \;|||\; push \to P$ resembles an infinite unary stack and therefore cannot be automatically transformed to a finite timed automaton. In order to prevent ill-formed definitions like this one, self-recursion within a *Parallel Composition* must not be used. Violations of this constraint can easily be detected: Whenever two locations are combined in a tuple (usually when translating a *Parallel Composition*), and one of them has been marked as an unresolved recursion stub, the recursion hierarchy could be tangled.

As an example of the transformation of recursive processes, consider a process structure like the following.

$R = \ldots Q \ldots P \ldots$
$Q = \ldots R \ldots$
$P = \ldots Q \ldots$

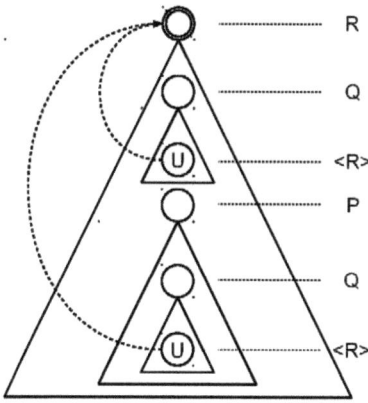

Our transformation first calculates the automaton representing Q and introduces a stub location $<R>$ referring to the process R, which is currently translated. Then, the automaton representing P is calculated including the automaton corresponding to Q. Finally, the translation of the timed automaton A_R representing R is finished and all stub locations <R> referring to R are connected to the initial location of A_R.

Within our prototypical implementation, we treat parameterized process variables by computing the reachable process variables (with respect to the respective valuations of the parameters) of the source model. Then, for each (reachable) "flat" process variable, a transformation of the corresponding process is performed.

6.1.3 Simulation and Automatic Verification in UPPAAL

We have implemented the transformation rules given above as a tool called TCSP2TA. It takes a set of Timed CSP processes as input fulfilling the assumptions above. Furthermore, a main process P to be translated is given. Then, the main process is translated in a compositional fashion to an intermediate representation of timed automata. If the main process refers to process names, the respective process bodies are translated according to the transformation of *Recursion* described above. Finally, our tool outputs an XML-file that UPPAAL can cope with. The UPPAAL model consists of two timed automata composed in parallel: The first automaton A_P corresponds to the main process, where all events a are declared as UPPAAL's receiving events $a?$. The second automaton is a simple control automaton that enables all possible events of A_P to happen. It consists of a single location that has for each event a a loop edge labeled with $a!$.

Within our framework, we translate finite instances of the abstract parameterized model and the concrete parameterized model S_k and N_k to their timed automata counterparts. The UPPAAL simulator can then be used to simulate the resulting models and to check whether they behave as ex-

6.1 Transformation from Timed CSP to Timed Automata

pected. If the simulation results in unintended behavior, the original model can be debugged according to these "wrong" behaviors.

The UPPAAL model checker can be used to verify properties of the instance models. One of these properties is that the resulting model is deadlock free, which is an important verification goal for embedded systems as considered in this thesis. This property can be encoded as the UPPAAL assertion A☐ not deadlock and then be automatically verified.

Properties given in our timed HML cannot directly be expressed in the UPPAAL specification language. We need to introduce further ghost variables and extend edges with corresponding updates to achieve this goal. Common (sub)properties that can be expressed in our logic are, for example, the following.

1. ☐ ([[a]]$_{=0}$ [[b]]$_{>d_u}$ ff)
2. ☐ ([[a]]$_{=0}$ [[b]]$_{<d_l}$ ff)
3. ☐ ([[a]]$_{=0}$ ⟨⟨b⟩⟩$_{\geq 0}$ tt)

The first two properties state that after performing event a, a possibly following event b occurs within the time bounds d_l to d_u. The third property states that after performing event a, it is possible to eventually perform event b.

For verifying the first and the second property, the idea is to keep track of the last performed visible event. When performing event a, a fresh clock x is reset to 0. When event b is enabled later, it is checked whether this clock is within the time bounds. To realize this idea, we give each event e occurring in the timed automaton a unique integer value val_e and extend the timed automata model according to the following steps.

- declare a global integer variable last that keeps track of the occurrence of the last (visible) event

Integration of Automatic Verification Tools

- declare a global clock x (which is reset to 0 when event a occurs)
- within the control automaton, extend the edges labeled with
  - a! by the additional updates `last = val_a` and `x = 0`
  - c! with c either visible or hidden but declared observable (this includes $c = b$) by the additional update `last = val_c`

Then, for proving the corresponding first and second formula correct, we let UPPAAL verify the following two assertions.

1. A☐ (last==val_a ∧ enabled(b) $\longrightarrow$ x ≤ d_u)
2. A☐ (last==val_a ∧ enabled(b) $\longrightarrow$ x ≥ d_l)

Unfortunately, UPPAAL does not directly support predicates to express that certain events are enabled. So, we take all locations $l_1, \ldots, l_n$ of the automaton A_P with an outgoing b?-edge. Then, enabled(b) is defined as A_P.l_1 ∨ ... ∨ A_P.l_n. Note that this suffices because in a translated automaton A_P, edges labeled with some event cannot contain guards, i.e., enabledness is equivalent with being in the respective locations with an outgoing b? edge. Also note that upper bounds can only be verified if either b is a hidden event in the original Timed CSP process or if it is contained in a *Timeout* or *Timed Interrupt* construction, otherwise the occurrence of b could be delayed for an arbitrarily long amount of time.

The third property is harder to express within the UPPAAL query language and cannot be expressed in general. We would like to express something like the CTL formula A☐ (a_occurred $\longrightarrow$ E◊ b_is_enabled). The problem with such a formula is, however, that nesting of path quantifiers is not supported in the UPPAAL query language. The only supported formula with a sort of nested path quantifiers is $\Phi \leadsto \Psi$. This formula is equivalent to the CTL formula A☐($\Phi \longrightarrow$ A◊ Ψ). In our case, we are only able to express a stronger property, which states that event b *must*

6.1 Transformation from Timed CSP to Timed Automata

eventually be enabled (and actually be performed) after event a occurred. This property is, however, still useful in many cases as sketched below. To verify this property on a translated Timed CSP process, the idea is again to keep track of the last visible event. Furthermore, it has to be ensured that after performing event a, on every path event b is finally performed but without another visible event occurring between a and b. To this end, we introduce a further boolean variable changed, which is set to true if after the first occurrence of the a event, another visible event but b is performed. Initially, changed is set to false. We extend the timed automata model by the following steps.

- declare a global integer variable last that keeps track of the occurrence of the last (visible) event

- declare a global boolean variable changed that keeps track of whether another event but b occurred after the first a event

- within the control automaton, extend the edges labeled with

    - a! by the additional updates: in the case that $a \neq b$, add the update changed = (last==val_a ? true : changed)[1]; in each case add the update last = val_a

    - b! by the additional update last = val_b

    - c! with $c \neq a$, $c \neq b$, and c either visible or hidden but observable by the additional updates changed = (last==val_a ? true : changed) and last = val_c

This means that after the event a occurs and thereafter an event $c \neq b$ is performed, changed is set to true and will never be reset to false again.

Then, we verify the following formula in UPPAAL.

last==val_a $\leadsto$ last==val_a $\wedge$ changed==false $\wedge$ enabled(b)

---
[1]This is equivalent to the update changed = (last==val_a or changed).

This formula expresses that after performing an *a* event, a location where *b* is enabled *must* be reached. Note that this formula additionally implicitly states that the enabled b event must also be performed. Otherwise, if another event but b was performed, changed would be set to true and a further occurrence of the a event could not be followed by a state where changed==false and enabled(b). As discussed above this UPPAAL formula expresses a stronger property than the original timed HML formula. However, it is still useful in many situations. The mechanism described above can especially be used to verify formulae of the form

$$\Box \ ([[a]]_{=0} \ (\langle\langle b_1 \rangle\rangle_{\geq 0} \ tt \ \lor \ \langle\langle b_2 \rangle\rangle_{\geq 0} \ tt \ \lor \ \ldots \lor \ \langle\langle b_n \rangle\rangle_{\geq 0} \ tt))$$

This expresses that when event *a* was performed, one of the events b_i must be performed thereafter. In combination with the first two timed HML formulae, it can, for example, be expressed that after event *a* was performed, some event b_i must be performed thereafter within the time bounds d_l to d_u.

6.1.4 Discussion

In this section, we have presented our adaptations and extensions of the transformation approach from Timed CSP to timed automata as originally given in [DHQ+08]. The main restrictions on processes that can be correctly transformed are that they need to be finite and that each *SKIP* in the original process should appear (as a subprocess) in the left process of a *Sequential Composition* to treat termination correctly. These are minor restrictions. Thus, we are able to transform a large class of finite Timed CSP processes to timed automata and can finally employ the UPPAAL tool suite for simulation, debugging, and verification. Furthermore, this transformation copes with our extended semantics of Timed CSP where hidden events can be declared as visible (not having any influence on other operators) and thus enables the verification with respect to internal behaviors of

6.2 Transformation from Timed CSP to Tock CSP

a Timed CSP process, which is crucial in the area of embedded real-time systems. Finally, we discussed how common timed HML subformulae can be verified using the UPPAAL model checker.

Although a correctness proof was given in [DHQ+08], the transformation contained flaws that we corrected by reformulating some of the original transformation rules. We have informally argued that our transformation is correct. However, especially in the treatment of *Recursion*, a hand-written correctness proof can easily contain flaws. In future work, we therefore plan to formalize timed automata and the given transformation rules in Isabelle/HOL. Using our formalization of Timed CSP, we aim at a formal and mechanical verification of the correctness of the presented transformation rules.

UPPAAL is well-suited to verify (timed) properties of translated Timed CSP processes. However, it lacks mechanisms to verify semantical refinement or semantical equivalence of two processes. To this end, we consider another transformation approach from Timed CSP to tock CSP such that the FDR2 refinement checker can be used for verification as explained in the following section.

6.2 Transformation from Timed CSP to Tock CSP

The aim of [Oua01] is to relate a continuous-time semantics and a special discrete-time semantics of Timed CSP in order to allow for the verification of continuous-time systems in terms of the verification of the discretely timed counterpart. One of the motivations of [Oua01] is to provide Timed CSP with a syntactic mapping $\Psi :: TCSP \Rightarrow tockCSP$. It is a compositional mapping that translates Timed CSP processes to discretely timed

Integration of Automatic Verification Tools

CSP processes, where the special event *tock* describes the passage of one time unit. In Section 6.2.1, we introduce assumptions on Timed CSP processes that are needed to ensure the transformation to be correct. Then, we present the transformation rules and our adaptations and extensions in Section 6.2.2. In Section 6.2.3, we present further assumptions that are needed such that the FDR2 refinement checker can cope with the translated processes. Furthermore, we show how the FDR2 refinement checker is used to verify timed traces equivalence in its tau-priority model. Finally, in Section 6.2.4, we summarize the presented transformation rules and discuss their applicability.

The overall transformation algorithm works as follows. If a process $P1 \oplus P2$ is to be translated where $\oplus$ is some Timed CSP operator, then first translate $P1$ and $P2$. These translated processes are combined using CSP operators in tock CSP according to the corresponding transformation rule. The transformation rules for all process operators are described below.

6.2.1 Assumptions

To ensure the correctness of the transformation rules below, we need to restrict the set of allowed processes.

- The presented mapping assumes that all timed operators in the Timed CSP process to be transformed are defined over natural numbers. This means, for example, that for a $WAIT(n)$ process being part of the overall process to be translated, it is assumed that $n \in \mathbb{N}$.

- For the *External Choice* process $P1 \square P2$, we assume that neither $P1$ nor $P2$ can terminate before communicating a visible event.

- For the *Timeout* process $P1 \stackrel{n}{\triangleright} P2$, we assume that $P1$ needs to perform at least one visible event before successfully terminating.

6.2 Transformation from Timed CSP to Tock CSP

- For the *Interrupt* process $P1 \triangle P2$, we have the restriction that $P2$ terminates exactly after the first performed visible event. Note that $P2$ is then basically of the form $WAIT(n); a \rightarrow SKIP$.

- For the *Timed Interrupt* process $P1 \triangle_n P2$, we impose the restriction that $P2$ has only the possibility to successfully terminate after some amount of time. So $P2$ can basically be of the form $WAIT(n)$.

- We make the assumption that a Timed CSP process to be transformed is finite.

Especially for *Interrupt* and *Timed Interrupt*, the restrictions are relatively strong. When giving the respective transformation rules below, we explain why it is hard to find more general transformation rules for these process operators. However, we also explain why these restrictions nevertheless allow for interesting processes to be transformed to tock CSP.

6.2.2 Transformation Rules

Under these assumptions, a Timed CSP process can be transformed into tock CSP according to the following rules.

Basic Processes The tock CSP representation of the Timed CSP process *STOP* is a recursive process performing infinitely many *tock* events.
$STOP_t \stackrel{def}{=} \Psi(STOP) = \mu X \bullet tock \rightarrow X$

The *SKIP* process also can perform arbitrarily many *tock* events and can always successfully terminate.
$SKIP_t \stackrel{def}{=} \Psi(SKIP) = \mu X \bullet tock \rightarrow X \square SKIP$

This gives a slightly different transformation compared to [Oua01]. There, a special operator $SKIP_t$ was introduced, which can let time arbitrarily advance as in our case but behaves like $STOP_t$ after successfully terminating. Such an operator can, however, not be expressed in standard CSP as supported by FDR2 because the only process introducing terminating events is $SKIP$. However, it evolves to $STOP$ and not $STOP_t$. This means that using our transformation after a terminating event no *tock* events may be performed anymore.

The *Prefix* operator $a \rightarrow P$ is realized by a process that allows arbitrarily many *tock* events to be performed and that is always able to perform the a event. When this happens, it behaves like the transformed process $\Psi(P)$.

$$\Psi(a \rightarrow P) = \mu X \bullet (a \rightarrow \Psi(P)) \square (tock \rightarrow X)$$

Menu Choice is treated in a similar way.

$$\Psi(x : A \rightarrow P_x) = \mu X \bullet (a : A \rightarrow \Psi(P_x)) \square (tock \rightarrow X)$$

The process *WAIT* is explicitly considered in the transformation. After performing n *tock* events it behaves like $SKIP_t$.

$$WAIT_t(n) \stackrel{def}{=} \Psi(WAIT(n)) = \underbrace{tock \rightarrow \ldots \rightarrow tock}_{n \text{ times}} \rightarrow SKIP_t$$

Internal Choice The Timed CSP process *Internal Choice* is defined in terms of CSP's *Internal Choice*, where the subprocesses are transformed accordingly.

$$\Psi(P1 \sqcap P2) = \Psi(P1) \sqcap \Psi(P2)$$

External Choice In [Oua01], the transformation for *External Choice* is proposed as follows.

$$\Psi(P1 \square P2) = \Psi(P1) \square_t \Psi(P2)$$

where

6.2 Transformation from Timed CSP to Tock CSP

$$P1' \square_t P2' = f_1^{-1}(f_2^{-1}((f_1(P1') \underset{\{tock\}}{\|} f_2(P2')) \underset{\Sigma_1 \cup \Sigma_2}{\|} (RUN_{\Sigma_1} \square RUN_{\Sigma_2})))$$

and for $i \in \{1,2\}$, we have

$\Sigma_i = \{i.a \mid a \in \Sigma\}$

$f_i :: \Sigma \cup \{tock\} \cup \Sigma_1 \cup \Sigma_2 \Rightarrow \Sigma \cup \{tock\} \cup \Sigma_1 \cup \Sigma_2$
$f_i(a) = $ **if** $a \in \Sigma$ **then** $i.a$ **else** a

$RUN_A = (a : A \rightarrow RUN_A)$

This definition first renames all events a of $P1'$ to $1.a$ and events a of $P2'$ to $2.a$ using the renaming functions f_1 and f_2, respectively. Then, these two processes are placed into a *Parallel Composition* and synchronize over *tock* events. When a visible event of either the first process or the second process is performed, the choice in $RUN_{\Sigma_1} \square RUN_{\Sigma_2}$ is resolved such that the events of the other process are blocked. Finally, events $1.a$ and $2.a$ are given their original names by f_1^{-1} and f_2^{-1}.

The semantics of Timed CSP given in [Oua01] defines termination of *Parallel Composition* in another way than in the original semantics of (Timed) CSP that is used by FDR2 and that we use in our formalization of Timed CSP. In [Oua01], a parallel process can successfully terminate if *one* of its subprocesses can. In the original semantics of (Timed) CSP, a parallel process can only terminate if *all* its subprocesses can. If, for example, in the above construction $P1'$ first communicates some event a and then terminates, the overall parallel composition cannot terminate, because $P2'$ cannot terminate. Therefore, we slightly change the definition above according to the original semantics as follows.

$\Psi(P1 \square P2) = \Psi(P1) \square'_t \Psi(P2)$

Integration of Automatic Verification Tools

$$P1' \square'_t P2' = f_1^{-1}(f_2^{-1}(((f_1(P1') \triangle trig_1 \to SKIP_t$$
$$\underset{\{tock\}}{\|} f_2(P2') \triangle trig_2 \to SKIP_t)$$
$$\underset{\Sigma_1 \cup \Sigma_2 \cup \{trig_1, trig_2\}}{\|} (CHOICE_1 \square CHOICE_2)) \setminus_t \{trig_1, trig_2\}))$$

$CHOICE_1 = a : \Sigma_1 \to trig_2 \to RUN_{\Sigma_1^\checkmark}$
$CHOICE_2 = a : \Sigma_2 \to trig_1 \to RUN_{\Sigma_2^\checkmark}$

$RUN_{A^\checkmark} = (a : A \to RUN_{A^\checkmark}) \square SKIP$

We have changed the original transformation rule by introducing the interrupts $trig_i \to SKIP_t$ using fresh events $trig_i$ and by introducing processes $CHOICE_i$ waiting for the first visible event of either $P1'$ or $P2'$. If, for example, $P1'$ performs a visible event the $CHOICE_1$ process communicates the $trig_2$ event thereafter such that the second process evolves to $SKIP_t$. If finally the respective process terminates, all processes within the *Parallel Composition* can terminate as well.

Remember that we assume in our transformation that both processes $P1$ and $P2$ need to communicate a visible event before terminating. Otherwise, the composed process would simply deadlock because neither $CHOICE_1$ nor $CHOICE_2$ are able to terminate initially. Even if these processes could initially terminate, then either $P1'$ or $P2'$ would be prevented from terminating because $trig_1$ or $trig_2$ would not be performed. This is a minor restriction because, for example, a Timed CSP process like $WAIT(5) \square P$ could be reformulated as $((WAIT(5); tmp \to SKIP) \square P) \setminus \{tmp\}$.

Sequential Composition *Sequential Composition* is translated by first translating both subprocesses and then combining them using a special urgent *Sequential Composition* operator. It expresses that *Sequential Composition* needs to be resolved as soon as possible, i.e., time must not advance as long as the first process is able to terminate. This means that if the first process of the *Sequential Composition* can terminate, it must not perform

6.2 Transformation from Timed CSP to Tock CSP

further *tock* steps. This treatment of the *Sequential Composition* operator is supported in the tau-priority model of FDR2, where each τ step gets priority over *tock* steps.

$$\Psi(P1;P2) = \Psi(P1);_t \Psi(P2)$$

Parallel Composition The transformation of *Parallel Composition* is given by the translation of the subprocesses, which are then composed in parallel and additionally synchronize over the *tock* event.

$$\Psi(P1 \underset{A}{\|} P2) = \Psi(P1) \underset{A\cup\{tock\}}{\|} \Psi(P2)$$

Timeout *Timeout* can be expressed in terms of the *WAIT* process using a fresh event *trig* by $P1 \overset{n}{\triangleright} P2 = (P1 \square (WAIT(n); trig \to P2)) \setminus \{trig\}$. Therefore, its transformation is simply given by

$$\Psi(P1 \overset{n}{\triangleright} P2) = (\Psi(P1) \square'_t (WAIT_t(n);_t trig \to \Psi(P2))) \setminus_t \{trig\}.$$

Like in the case of *External Choice*, we assume that *P1* needs to perform a visible event before successfully terminating. This restriction does not need to be imposed on *P2* because of the introduced *trig* event, which resolves the underlying *External Choice*.

Hiding As in the case of *Sequential Composition*, *Hiding* is translated using a special urgent *Hiding* operator. It means that *tock* must not be performed when an event to be hidden can be performed. Again, this operator is available in the tau-priority model of FDR2.

$$\Psi(P1 \setminus A) = \Psi(P1) \setminus_t A$$

Interrupt In [Oua01], the process operators *Interrupt* and *Timed Interrupt* have not been considered. Inspired by the transformation rule of *Ex-*

Integration of Automatic Verification Tools

ternal Choice, we developed the following transformation rule for *Interrupt*.

$$\Psi(P1 \triangle P2) = \Psi(P1) \triangle_t \Psi(P2)$$

$$P1' \triangle_t P2' = f_1^{-1}(f_2^{-1}(((f_1(P1') \triangle trig \to SKIP_t \underset{\{tock\}}{\|} f_2(P2') \triangle SKIP)$$

$$\underset{\Sigma_1 \cup \Sigma_2 \cup \{trig\}}{\|} RUN_1 \square IR)) \setminus_t \{trig\}))$$

$$RUN_1 = (a : \Sigma_1 \to (RUN_1 \square IR)) \square SKIP$$

$$IR = a : \Sigma_2 \to trig \to RUN_{\Sigma_2^{\checkmark}}$$

As in the case of *External Choice*, we have the assumption that $P2'$ can only terminate after performing some visible event. If $P1'$ performs a visible event, the *Interrupt* is not resolved in favor of it and therefore we do not have the *trig* interrupt on the right hand side. If $P1'$ performs a *tock* event, it needs to be synchronized with a *tock* event performed by $P2'$. Finally, if $P1'$ terminates, it synchronizes with $f2(P2') \triangle SKIP$ and with RUN_1.

For process $P2'$, there is a problem when it performs a visible event. Then, process *IR* performs the *trig* event to interrupt $f_1(P1')$. However, the term $\triangle SKIP$ remains. Therefore, the overall process can terminate now, even if $P2'$ cannot. Replacing the term $\triangle SKIP$ by $\square SKIP$ does not solve this problem because then *tock* events would resolve the choice. We believe that it is not (directly) possible to define a general *Interrupt* operator in tock CSP.

Although not having found a general transformation rule for *Interrupt*, the above transformation rule transforms $P1 \triangle P2$ correctly if the second process terminates after performing the first visible event. With the additional assumption that it can only terminate after performing a visible event, we basically allow processes of the form $WAIT(n); x : A \to SKIP$

6.2 Transformation from Timed CSP to Tock CSP

for *P2*. This is a quite restricted form of the *Interrupt* process but it is nevertheless still useful in many situations.

For example a police station could be modeled where an emergency event interrupts the office work. Then, the emergency is handled and after this the process starts over again.

$Police = (Office \triangle emergency \rightarrow SKIP); Emergency\_Handling; Police$

Timed Interrupt We establish the transformation of *Timed Interrupt* by a characterization in terms of the *Interrupt* operator and the *WAIT* process using a fresh event *trig*.

$\Psi(P1 \triangle_n P2) = \Psi(P1) \triangle_t (WAIT_t(n);_t trig \rightarrow \Psi(P2)) \setminus_t \{trig\}$

Due to the restrictions with respect to *Interrupt*, *P2* must be restricted to be basically of the form *WAIT*(*k*) (with *SKIP* being a special case since *SKIP* is equivalent to *WAIT*(0)). However, this operator is, for example, useful to restrict the time of operation of a process:

$(PARTY \triangle_5 SKIP); goto\_bed \rightarrow STOP.$

Recursion If a set of (possibly recursive) process definitions $R_i = P_i$ is given, then all bodies are translated according to the transformation rules given above: $R_i = \Psi(P_i)$. Process variables that occur in P_i are not changed by Ψ.

Like in the case of TCSP2TA, we treat parameterized process variables by computing the reachable "flat" process variables. Then, for each such process variable, a transformation of the corresponding process is performed.

Integration of Automatic Verification Tools

6.2.3 Automatic Verification with FDR2

We have implemented the transformation rules as given above as a tool called TCSP2tockCSP. It takes a set of Timed CSP processes as input and translates them according to the transformation rules. The output is a text file that FDR2 can interpret. However, using the presented transformation rules, the state space of a translated process can be infinite even if it was finite for the original Timed CSP process as described below. Therefore, we need to further restrict the Timed CSP processes to be transformed as explained in the following.

Transformation Assumptions for Automatic Verification in FDR2

- Within the translation of *External Choice* there are two severe problems in the context of automatic verification using FDR2: Consider the example process $R = a \to R \square b \to R$. This process is translated to a tock CSP process with infinite state space according to the transformation rule for *External Choice* above. This is because after each recursive call the overall (transformed) process grows further as it does not get rid of the *Parallel Composition* operators. There are two possibilities to circumvent this problem in special cases.

  The first case is that the subprocesses of the *External Choice* are either of the form $a \to P1$ or $x : A \to P_x$. Then, the transformation can simply be given by introducing an additional loop for *tock* events and by replacing the reached process after the occurrence of some visible event with their respective translations. We have enhanced our transformation tool by this simple analysis. For example, the process

6.2 Transformation from Timed CSP to Tock CSP

$R = (a \rightarrow P1) \square (x : A \rightarrow P_x) \square (b \rightarrow P3)$

is translated to

$R = (tock \rightarrow R) \square (a \rightarrow \Psi(P1)) \square (x : A \rightarrow \Psi(P_x)) \square (b \rightarrow \Psi(P3))$.

The second case is that in a recursive process like $R = \ldots P \square Q \ldots$, neither P nor Q refer to R but P and Q are themselves possibly recursive. This poses no difficulties to FDR2 in the translated version.

- The transformation of *Timeout* $P1 \overset{n}{\triangleright} P2$ is based on the transformation of *External Choice*. The second process waits for the timeout to elapse and then resolves the *External Choice* if the first process did not yet communicate a visible event in the meantime. According to the discussion concerning *External Choice* above, we assume that within a recursive process definition $R = \ldots P1 \overset{n}{\triangleright} P2 \ldots$, neither $P1$ nor $P2$ refer to R.

- For the translation of *Interrupt* and *Timed Interrupt*, we impose the restrictions that whenever they occur in a recursion

$R = \ldots P1 \triangle P2 \ldots$ or
$R = \ldots P1 \triangle_n P2 \ldots$,

neither $P1$ nor $P2$ refer to R. Note that these restrictions are already implicitly contained in the restrictions in Section 6.2.1. In the case of *Interrupt*, $P2$ was assumed to terminate after exactly the first visible event. In the case of *Timed Interrupt*, $P2$ was assumed to terminate after some time. So in both cases, $P2$ cannot refer to R. If on the other hand $P1$ referred to R, a process with an infinite state space would be introduced. Our transformation, however, assumes the Timed CSP processes, which are transformed, to be finite.

Within our framework, we want to show that instances of an abstract and a concrete parameterized system are semantically equivalent. The current version of the FDR2 refinement checker only supports traces re-

Integration of Automatic Verification Tools

finement in the tau-priority model. Therefore, we can employ the FDR2 refinement checker to show timed traces equivalence of instances of parameterized real-time systems. As shown in Section 5.2.5, weak timed bisimulation implies equality of timed traces. Thus, we get an evidence that the comprehensive bisimulation-based verification is possible in the following sense: Assume that the timed traces equality check using the FDR2 refinement checker for system instances was not successful and nevertheless the bisimulation proof could be established. Then, because the formerly considered instances would be part of the bisimulation relation, this would imply timed traces equivalence for these instances, which leads to a contradiction. So by establishing the timed traces equivalence we cannot conclude bisimilarity but we get a certain evidence that the proof for this is possible.

To show timed traces equivalence between instances of the abstract and concrete parameterized real-time systems, we show the following two assertions correct using FDR2.

```
assert A_sys_k [= C_Sys_k :[tau priority over]:{tock}
assert C_sys_k [= A_Sys_k :[tau priority over]:{tock}
```

If one of these checks is not successful, FDR2 provides a counterexample that can be used to debug the considered models. Furthermore, simulation tools such as ProB [LF08] or ProBE [PRO03] can be used to simulate the counterexamples. However, as neither of these tools support the tau-priority model, τ events have to be performed manually in favor of possible *tock* events.

6.2.4 Discussion

In this section, we have presented our adaptations and extensions of the transformation approach from Timed CSP to tock CSP as given in [Oua01].

6.2 Transformation from Timed CSP to Tock CSP

There are several restrictions on the Timed CSP processes such that they can be transformed correctly. This is mainly due to problems concerning termination in parallel processes and from recursion within parallel processes. However, we are still able to transform a large class of finite Timed CSP processes to tock CSP and can finally employ the FDR2 refinement checker for verification and debugging of processes.

In the context of our verification framework, special care has to be taken concerning *Hiding* of observable events. Furthermore, the soundness of the discretization approach of [Oua01] requires the verification of further proof obligations. However, these proof obligations can be omitted for our purposes as explained below.

Hiding of Observable Events Within our verification framework, the final verification goal is to prove that the concrete parameterized real-time system $N_n \setminus A$, where A may contain events declared as observable, satisfies given requirements R_n. To this end, we aim at showing, among others, that N_n is weak timed bisimilar to an abstract parameterized system S_n. The presented transformation from Timed CSP to tock CSP is used to show that instances N_k and S_k are semantically equivalent using FDR2. Our notion of weak timed bisimulation explicitly considers hidden but observable events. However, hidden events in FDR2 are all mapped to the indistinguishable internal event τ. Therefore, we need to make the assumption that neither in N_k nor in S_k observable events are already hidden to get a correct verification result for the instance verification.

Closure Under Inverse Digitization In [Oua01], a discretization approach for Timed CSP is presented. It is proved that the discretization gives correct verification results with respect to the refinement $S \sqsubseteq P$ where the involved processes are interpreted over a continuous-time semantics if the refinement $S \sqsubseteq P$ holds in the discrete setting and the denotational seman-

Integration of Automatic Verification Tools

tics of S is *closed under inverse digitization*. This intuitively means that from discrete-time interpretations of Timed CSP processes, the continuous-time interpretation can be inferred to a certain degree. In [OW03], it has been shown that certain classes of specification processes satisfy the property of being closed under inverse digitization. In general, however, this cannot be decided.

In our work, we do not explicitly consider closure under inverse digitization as this would lead to additional proof obligations in the denotational semantics of Timed CSP, which we did not (yet) formalize in the Isabelle/HOL theorem prover. Furthermore, concerning the integration of FDR2 into our verification framework, to be able to verify this property would in the best case give some greater evidence that the final bisimulation proof can principally be performed. However, the verification in FDR2 still gives an evidence independent of whether closure under inverse digitization is satisfied or not: If two Timed CSP processes P and Q are weak timed bisimilar, then they have equal timed traces as shown in Section 5.2.5. Then, they also have equal discretely timed traces as shown in [Oua01]. The latter verification goal can be achieved using the FDR2 refinement checker. So if this check fails, the weak timed bisimulation proof is not possible. In summary, this means that performing the corresponding equivalence proof in FDR2 still gives an evidence that the bisimulation proof is possible, even without showing closure under inverse digitization of P and Q. Remember that we aim at finding bugs in the considered parameterized real-time systems using the FDR2 refinement checker. As the verification of (in the worst case) discrete timed traces equivalence still copes with this issue, not having absolute assurance concerning continuous timed traces is not a problem.

6.3 Summary

In this chapter, we have presented our transformations from Timed CSP to timed automata and from Timed CSP to tock CSP. They are based on approaches presented in [DHQ$^+$08] and [Oua01], which we have adapted and extended to fit the (original) semantics of Timed CSP and to support a larger class of processes that can be transformed.

In [DHQ$^+$08], the transformation rules concerning *External Choice* and *Timeout* contain flaws, which we have corrected as described above. Furthermore, we have developed transformation rules for a general form of the *Interrupt* operator and *Hiding*, which were not considered in [DHQ$^+$08]. We have implemented our transformation rules and have shown how the resulting timed automata model corresponding to a Timed CSP process can be safely extended to verify common timed Hennessy-Milner logic properties using the UPPAAL model checker.

In [Oua01], a different semantics concerning termination of parallel processes with respect to the original semantics of (Timed) CSP is used. Because it differs with respect to the semantics of FDR2, we have adapted the transformation rule of *External Choice*. This, however, requires additional assumptions concerning the Timed CSP processes to be transformed. The *Interrupt* and *Timed Interrupt* processes are not considered in [Oua01]. Therefore, we have developed transformation rules for these operators. It has turned out that the handling of termination within the transformation of *Interrupt* is quite complicated. Therefore, we only permit a very restricted form of *Interrupt* and *Timed Interrupt*. However, we have argued that this restricted form is still useful in many situations. We have implemented our transformation rules and have shown how the FDR2 refinement checker can be employed for the verification of instances within our verification framework.

Integration of Automatic Verification Tools

Up to this point, we have presented our framework for the mechanical verification of parameterized real-time systems in Chapter 4, which is based on our formalization of Timed CSP and the verification techniques of bisimulations and coinductive invariants with respect to timed HML formulae (Chapter 5). In this chapter, we have presented transformations from Timed CSP to timed automata and tock CSP, which enhance our framework by the possibility to employ automatic verification tools for simulation, debugging, and instance verification. In the next chapter, we present the application of our overall framework to our case study of a simplified, but yet illustrative, model of a real-time operating system scheduler.

7 Case Study - A Real-Time Operating System Scheduler

In this chapter, we report on the application of our verification framework to the case study of a real-time operating system scheduler. With our framework, we are able to verify possibly infinite parameterized real-time systems comprehensively. Furthermore, we enable the automatic simulation and verification of finite instances by employing our transformation engines from Timed CSP to automatically analyzable languages. Our case study shows the applicability of our verification framework. It is an implicitly infinite parameterized system for which timing and functional properties are verified. In Section 7.1, we present a model of the parameterized real-time scheduling system and present timed HML properties that describe the requirements of this system. Furthermore, we provide an abstract version of the parameterized real-time system that is structurally simpler for later verification purposes. In Section 7.2, we present how we use the UPPAAL model checker and the FDR2 refinement checker to verify small instances of the parameterized scheduler after transforming

Case Study - A Real-Time Operating System Scheduler

them to timed automata and tock CSP, respectively. The phase of comprehensive mechanical verification of the scheduling system is explained in Section 7.3. This section is divided into two parts: First, we explain how the bisimulation proof is established, with which we show that the concrete and the abstract parameterized real-time systems are weak timed bisimilar for all network sizes. Second, we sketch how the requirements given as timed HML properties are verified on the abstract parameterized scheduling system. These two verification results are integrated to achieve the final verification goal: The concrete parameterized scheduling model satisfies the given requirements for all network sizes. This chapter closes with a short summary concerning the presented case study in Section 7.4.

7.1 Timed CSP Model of a Real-Time Scheduler

The real-time operating system BOSS [MRH05] is the motivating case study in the VATES project. It was developed at Fraunhofer FIRST and is deployed in, for example, the DLR satellite BIRD designed for early fire detection. In this section, we develop a simplified model of BOSS's single-core scheduler in Timed CSP and verify it. The scheduler process runs in parallel with arbitrarily many threads that are controlled by it. Each thread is characterized by a name (i) and a rank ($rank\ i$) where $rank\ i < rank\ j$ means that thread i has higher priority than thread j. We assume that each of the threads, when having finished its computations, is to be rescheduled after all other threads have finished their computations. We further assume that the priorities of the threads in the system are fixed and that each thread i can be given a maximal amount of time $tt\_p\ i$ for which it needs control over the computing resources in one scheduling cycle.

7.1 Timed CSP Model of a Real-Time Scheduler

The scheduler first initializes the context of a thread and then gives control to the thread with the highest priority. After this it waits for the thread to yield its control or may preempt it. If the thread yields its control, it may only be rescheduled after all other threads in the system have finished their computations. When it is preempted, another thread of the same priority is given control. The preempted thread is inserted as the last one of its priority. When a thread has been started, it can compute something, send a message to another thread, yield its control (thus finishing its computation) or it can be preempted by the scheduler.

7.1.1 Formalizing the Model

Events For describing the events of our model, we define the datatype ev where the constructors have the meaning that a thread can be started, yield its control and that it can be preempted. Furthermore, a thread can do something with a received message and threads can exchange messages. Finally, the scheduling cycle is restarted if all threads finished their computations.

```
datatype ev = start_ev "id" | yield_ev "id" | preempt_ev "id"
            | do_ev "id × mess" | send_ev "id × id × mess"
            | restart_ev
```

Threads of the system are identified by id, which is realized as a type synonym for nat. To give the events their usual look as in (Timed) CSP, we provide syntax abbreviations such that, for example, a send event can be written as send°i°j°m instead of send_ev(i,j,m).

Process Variables For describing the parameterized recursive processes of the scheduling system, we define five kinds of process variables using Isabelle's datatype mechanism.

Case Study - A Real-Time Operating System Scheduler

```
datatype procn = Sch "id list × id list"
               | Thread "id × nat × mess × nat"
               | Thread_body "id × nat × mess × nat × bool"
               | Inv "nat × mess list × nat list"
               | Inv_body "id × nat × mess list × nat list × bool"
```

- Sch(l,init) denotes the scheduler, which executes the threads corresponding to the identifiers in the (sorted) list l and starts over again with init when all threads have been executed, i.e., l is empty.

- Thread(i,n,m,t) denotes the thread with identifier i, which depends on the network size n, is working on the data of message m, and has t time units left to run.

- Thread_body(i,n,m,t,some) denotes the thread with identifier i that has currently control over the computing resources and records whether it has already done something or has sent a message to another thread since it was started.

The following process variables are used to model the abstract parameterized system as described below.

- Inv(n,lm,lt) denotes a network of threads in one single process and keeps track of the size n of the network, memorizes the messages each thread is working on in lm and memorizes the residual times of each thread in the system in lt.

- Inv_body(i,n,lm,lt,some) denotes the state of the network where the thread with identifier i has been started and records whether this thread has done something or has sent a message since has been started.

Timed CSP Model The scheduler model works on abstract static parameters. Therefore it is defined in the context of the locale prio_sched.

7.1 Timed CSP Model of a Real-Time Scheduler

```
locale prio_sched =
  fixes rank::    id ⇒ nat
    and tt_p::    id ⇒ nat
    and t_init::  real
assumes t_init_pos: "t_init > 0"
```

The rank function models the (static) priorities of the threads. In the case of rank i < rank j, the priority of the thread with identifier i is higher than the priority of the thread with identifier j. The function tt_p maps a worst case execution time to each thread of the system. Finally, t_init denotes the time that the scheduler needs to change the context of a thread. It is assumed that this value is greater than 0.

Within this locale, we are now able to define the process variable assignment asg giving the introduced process variables their meaning as follows. The functions hd and tl give the head or the tail of a list, respectively.

```
fun asg:: procn ⇒ (procn,ev)Process"
where
 "asg ( Sch(l,init) ) =
  if l=[] then restart → <Sch(init,init)> else
  WAIT(t_init) ;; start°hd(l) →
    (yield°hd(l) → <Sch(tl(l),init)>
      □
    preempt°hd(l) → <Sch(insort_key2 rank hd(l) tl(l),init)>)"

| "asg ( Thread(i,n,m,t) ) =
    start°i → <Thread_body(i,n,m,t,False)>
    □ send_ev?x:{(j,i,nm). j>0 ∧ j≤n ∧ rank j≤rank i ∧ i≠j}
                       → <Thread(i,n,thrd(x),t)>"

| "asg ( Thread_body(i,n,m,t,some) ) =
    (t>0) && (do°i°m → WAIT(1) ;; <Thread_body(i,n,m,t−1,True)>)
    □ (t>1) && (send_ev?x:{(i,j,nm). j>0∧j≤n ∧ i≠j ∧ rank i≤rank j}
                       → WAIT(2) ;; <Thread_body(i,n,m,t−2,True)>)
    □ (some) && (yield°i → <Thread(i,n,m,tt_p(i))>)
    □ (some ∧ t>0) && (preempt°i → <Thread(i,n,m,t)>)"
    ...
```

Case Study - A Real-Time Operating System Scheduler

In the case that the schedule list is empty, the scheduler is restarted with the initial schedule list `init`. Otherwise, the scheduler first initializes the context of the next thread to be scheduled, which takes `t_init` time units. After that, the first thread of the schedule list `l` is started and the scheduler waits for it to yield control again or that it is preempted. If a thread yields control, it will not be considered in the current scheduling cycle until the scheduler is restarted again. If it is preempted, the respective thread identifier is sorted into the schedule list just after all other thread identifiers with the same priority by the function `insert_key2`.

A thread can either be started or receive a message from a thread with higher priority (i.e., lower rank). If it is started, `some` is set to `False`, denoting that it has not done anything yet. If enough time is left, the thread (`thread_body`) can do something or send a message to another thread of the network with lower priority. Furthermore, it can yield control or be preempted after it has done something or sent a message.

The abstract network `Inv`, which models all threads of the system at once, is defined as follows.

```
   ...
| "asg  ( Inv(n,lm,lt)  )  = 
     (start_ev?i:{p.  p>0 ∧ p≤s}  →  <Inv_body(i,n,lm,lt,False)>)"

| "asg  ( Inv_body(i,n,lm,lt,some)  )  = (let  t=lt!(i−1) ; m=lm!(i−1) in
  (t>0) && (do°i°m  →  WAIT(1);;<Inv_body(i,n,lm,lt[i−1:=t−1],True)>)
  □
  (t>1) && (send_ev?x:{(i,j,nm). j>0∧j≤n ∧ i≠j ∧ rank i≤rank j}
      →  (WAIT(2)  ;;
         <Inv_body(i,n,lm[scnd(x)−1:=thrd(x)],lt[i−1:=t−2],True)>))
  □
  (some) && (yield°i  →  <Inv(n,lm,lt[i−1:=tt_p(i)])>)
  □
  (some ∧ t>0) && (preempt°i  →  <Inv(n,lm,lt)>))"
```

Initially, some process within the network can be started. If some thread has been started, the abstract network mimics the possible steps of the

7.1 Timed CSP Model of a Real-Time Scheduler

started thread as described above. To memorize the messages and the residual times in the system, the lists `lm` and `lt` are updated accordingly. As it is assumed that the identifiers of threads go from 1 to n in a scheduling system with n threads, updates of thread identifier `i` in `lm` and `lt` refer to position `i-1`.

The Assembled System The scheduler communicates over the event set A, consisting of `start`, `yield` and `preempt` events, with the threads of the system. Furthermore, for describing the network of threads, we define their alphabet `alph` explicitly, where they can communicate over shared events. The set of shared events consists of `send` events. This means that two threads can only exchange messages synchronously and are independent in all other steps.

```
definition A:: ev set
where A ≡ (start_ev 'UNIV) ∪ (yield_ev 'UNIV) ∪ (preempt_ev 'UNIV)

definition alph:: nat ⇒ ev set
where
  alph i ≡ send_ev '{(i1,i2,m). i1=i ∨ i2=i}
         ∪ do_ev '{(i1,m). i1=i}
         ∪ {start°i , yield°i, preempt°i}
```

The concrete parameterized scheduling system consists of the scheduler process and a network of arbitrarily many threads (depending on the input parameter n). In the initial state of the overall system, the scheduler and all involved threads are in their respective initial state. The schedule list consists of all thread identifiers of the threads, which are initially sorted according to their ranks.

```
definition C_sys:: nat ⇒ (procn,ev)Process
where
  C_sys n ≡ let prio_list = sort_key rank [[1..n]] in
        <Sch([],prio_list)>
              |[A]|
        || n* [alph] (λ i. <Thread(i,n,ml,tt_p(i))>)
```

183

Case Study - A Real-Time Operating System Scheduler

The abstract parameterized scheduling system consists of the scheduler process and the abstract network Inv, modeling all threads of the network at once.

```
definition A_sys:: nat ⇒ (procn,ev)Process
where
  A_sys n ≡ let prio_list = sort_key rank [[1..n]];
                lm_init = replicate n ml ;
                lt_init = map tt_p [[1..n]]
            in <Sch([],prio_list)> |[A]| <Inv(n,lm_init,lt_init)>
```

Until now, we have modeled the syntax of the involved Timed CSP processes in the scheduling system. To make the operational semantics available for them, we declare prio_sched to be a sublocale of tcsp using asg as the process variable assignment and start_ev ' UNIV as the set of observable hidden events.

```
sublocale prio_sched ⊆ sched: tcsp "asg" "start_ev 'UNIV"
```

Verification Goals The overall verification goal is to show that the concrete parameterized scheduling system invariantly satisfies the requirements defined in the next subsection. To this end, we hide all system events except restart to achieve urgency for them. Only start events are hidden *and* observable, as introduced by the sublocale above, which interprets the scheduling system within Timed CSP.

```
definition HS:: ev set
where HS ≡ UNIV − {restart}
```

Subsequently, we first show that the concrete and abstract parameterized scheduling systems (without any events hidden) are weak timed bisimilar.

(1) $\forall n > 0.\ (C\_sys\ n, A\_sys\ n) \in weak\_timed\_bisimilar$

Second, we show that the abstract parameterized scheduling system (with hidden events) invariantly satisfies the requirements.

7.1 Timed CSP Model of a Real-Time Scheduler

(2) $\forall n > 0.\ A\_sys\ n \setminus HS \models \Box R\ n$

Then, we can finally conclude that also the concrete scheduling system invariantly satisfies the given requirements

$\forall n > 0.\ C\_sys\ n \setminus HS \models \Box R\ n.$

From (1) and the congruence property of weak timed bisimulation, we can conclude that also the processes $C\_sys\ n \setminus HS$ and $A\_sys\ n \setminus HS$ are weak timed bisimilar. Then, by (2) and the fact that timed HML properties are preserved by weak timed bisimulation, we can conclude that also the concrete scheduling system satisfies the given requirements.

This two-step-approach has the advantage that, by showing the semantical equivalence of $C\_sys\ n$ and $A\_sys\ n$, we can concentrate on aspects concerning the verification with respect to parameterization. As a result, we get rid of the unbounded explicit network as present in C_sys. In the verification of the requirements of $A\_sys\ n \setminus HS$, we concentrate on aspects concerning timed and functional verification of the system and do not need to explicitly concentrate on the unbounded network of threads in the scheduling system.

In the following subsection, we formalize the requirements, which are to be verified for our concrete parameterized real-time scheduling system.

7.1.2 Formalizing the Requirements

We aim at verifying two requirements for the scheduling system of arbitrary network sizes n>0. The first requirement is used to show correctness with respect to timing properties, while the second is used to show a functional correctness requirement. We have chosen these two properties as examples as they demonstrate that our framework copes with both classes of properties.

Case Study - A Real-Time Operating System Scheduler

The first requirement R1 expresses that whenever a thread was started, it does get the control over the computing resources neither for more than d_u nor for less than d_l time units.

```
definition R1:: nat ⇒ (ev eventplus)formula
where R1 n ≡
  ⋀[1,n] (λ i. [[ τ_start°i ]]_≤0
            (⋀[1,n] (λ j. [[ τ_start°j ]]_<d_l  ff
                        ∧_f
                        [[ τ_start°j ]]_>d_u  ff
            )
          )
  )
```

It states that when some thread with identifier i was started, the next thread j is started neither before d_l nor after d_u time units. This means that thread j can only be started somewhere in between these time bounds. The values of d_l and d_u are given by t_init + 1 and t_init + tt_p(i), respectively.

The second property expresses that the scheduler is supposed to respect the priorities of the threads.

```
definition R2:: nat ⇒ (ev eventplus)formula
where R2 n ≡
  ⋀[1,n] (λ i. [[ τ_start°i ]]_≤0
            (⋀[1,n] (λ j. if (rank j < rank i)
                         then [[ τ_start°j ]]_≥0  ff
                         else tt
            )
          )
  )
```

After a thread with identifier i was started, it is not possible to start a thread with identifier j thereafter that has a lower rank and therefore a higher priority than i because threads with higher priority must already have been completed in the current scheduling cycle. Note that this property is basically the reason for including the event restart in our model. When the scheduler gives control to the last thread in its schedule list, the scheduler restarts with the initial list. If we did not model this by communicating a restart event, which is either not hidden or when hidden then observable, we were not able to show this property.

Finally, the overall requirement is simply the conjunction of the first and second sub-requirements.

```
definition R :: nat ⇒ (ev eventplus) formula
where R n ≡ R1 n ∧_f R2 n
```

Before performing the proofs in the Isabelle/HOL theorem prover, we have verified instances of the parameterized scheduling system using the UPPAAL model checker and the FDR2 refinement checker as explained in the next section. This phase helped a lot when we modeled the scheduling system in order to debug it.

7.2 Instance Verification using UPPAAL and FDR2

In the instance verification phase, we have considered the concrete network sizes k=2 and k=3. Furthermore, we have taken tt_p(i) = 4-i, t_init = 2 and experimented with different rank functions. In the instance models with two threads, we have defined the rank function such that the thread with identifier 1 has lower rank than the thread with identifier 2. In the instance models with three threads, we have defined the rank of the thread with identifier 1 to have the lowest rank and experimented with the case where thread 2 has lower rank than thread 3 and the case where thread 2 and thread 3 have equal ranks.

Case Study - A Real-Time Operating System Scheduler

7.2.1 Model Checking in UPPAAL

We have translated the instances (C_sys k)\HS and (A_sys k)\HS to UPPAAL timed automata using our TCSP2TA transformation engine. The only visible events are the start events and the restart event. We have extended the control automaton as formerly described in Section 6.1: The global variable last is used to record the last visible event. We have defined the identifier val_a to be i for the hidden but observable events _hidden_start_i and to be 0 for the event restart. The clock x is reset when performing the start events.

In order to show that the lower bound for consecutive start events is t_init + 1 = 3, we have verified the following formula in UPPAAL.

$A\square$ (last$\neq$0 $\wedge$ (enabled(_hidden_start_1)
    $\vee$ enabled(_hidden_start_2)
    $\vee$ enabled(_hidden_start_3)) $\longrightarrow$ x $\geq$ 3)

For upper bounds on the time between consecutive start events, we checked the following formulae.

$A\square$ (last==1 $\wedge$ (enabled(_hidden_start_1)
    $\vee$ enabled(_hidden_start_2)
    $\vee$ enabled(_hidden_start_3)) $\longrightarrow$ x $\leq$ 5)

$A\square$ (last==2 $\wedge$ (enabled(_hidden_start_1)
    $\vee$ enabled(_hidden_start_2)
    $\vee$ enabled(_hidden_start_3)) $\longrightarrow$ x $\leq$ 4)

$A\square$ (last==3 $\wedge$ (enabled(_hidden_start_1)
    $\vee$ enabled(_hidden_start_2)
    $\vee$ enabled(_hidden_start_3)) $\longrightarrow$ x $\leq$ 3)

Note that in the models with k=2, enabled(_hidden_start_3) is false and the case that last==3 cannot occur.

7.2 Instance Verification using UPPAAL and FDR2

For verifying that the scheduler respects priorities, we have introduced two global variables p1 and p2, which record the ranks of the last two start events. To this end, edges labeled with _hidden_start_i perform the updates p1=p2, p2=rank_i and the restart edge performs the updates p1=0, p2=0. Then, we have checked the formula

A$\Box$ p1 $\leq$ p2.

This means that if first _hidden_start_i and then _hidden_start_j is performed, the variable p1 is equal to rank_i and p2 is equal to rank_j and the property that rank_i $\leq$ rank_j is shown by the assertion above. If the scheduling is restarted, the variables p1 and p2 are reset to 0. This treatment is sound because the corresponding timed HML property only considers consecutive start events with no other (visible) events occurring in between.

Using these checks, we have verified that the requirements hold for the considered instances of the parameterized scheduling system. Thereby, we get the evidence that the comprehensive proof concerning the verification of the requirements for the abstract parameterized scheduling system can be established. Furthermore, as both the instances of the concrete and the abstract parameterized scheduling system satisfy the given requirements and the performed (successful) simulation results using the UPPAAL simulator already give some evidence that also the comprehensive bisimulation proof in Isabelle/HOL can succeed. To make the semantical relationship between instances of the concrete and abstract parameterized scheduling system more explicit, we perform the semantical equivalence proof using the FDR2 refinement checker as explained in the following.

Case Study - A Real-Time Operating System Scheduler

7.2.2 Refinement Checking in FDR2

We have translated the considered instances `C_sys k` and `A_sys k` to tock CSP using our TCSP2tockCSP transformation engine. Then, we have verified their semantical equivalence in the tau-priority model with respect to the traces model by successfully checking the following assertions in FDR2.

```
assert A_sys_k [= C_sys_k :[tau priority over]:{tock}
assert C_sys_k [= A_sys_k :[tau priority over]:{tock}
```

for k ∈ {2,3}.

This verification result means that both `C_sys_k` and `A_sys_k` produce the same timed traces. As explained in Section 6.2, this result gives some useful evidence that the comprehensive bisimulation proof in Isabelle/HOL is possible.

7.2.3 Transformation Times

All checks above have been performed in a few seconds in the UPPAAL model checker and the FDR2 refinement checker. Also, the transformations of the considered (relatively small) instance models have taken only a few seconds, except in the case where we translated `C_sys 3\HS` to timed automata. This already has taken about one minute, i.e., the scalability of the transformation is limited. The limiting factors are currently the number of parallel processes and the (reachable) range of parameters used in process variables, like imposed by the `tt_p` function, within these instance models. The latter can introduce many process variables whose number grows exponentially with the range of parameters. For each reachable process variable (all reachable valuations), currently a separate transformation of the corresponding process is performed. Additionally, in TCSP2TA, the

7.2 Instance Verification using UPPAAL and FDR2

transformation of the parallel composition involves the computation of a syntactical cross product. If the number of locations of the involved automata becomes too large, the translation takes a lot of time in practice. This problem does not occur in TCSP2tockCSP but the number of reachable process variables is also a limiting factor there. Thus, a tt_p function that gives the threads too much computation time can lead to very many process variables, which have to be considered in the transformation engines separately.

We additionally checked the transformation times for the cases k=4 to k=6. In the case of C_sys k, this already exceeds the memory bounds for TCSP2TA but is still possible in below one second in TCSP2tockCSP for all cases. For translating A_sys k in TCSP2TA, the transformation has taken 5 seconds for k=4, about one minute for k=5 and exceeds memory bounds for k=6. For translating A_sys k in TCSP2tockCSP, the transformation has taken 4 seconds for k=4, about one minute for k=5 and about 3 minutes for k=6. All transformations have been performed on a general purpose notebook. This shows that proving properties of instances of the parameterized real-time system in UPPAAL should be performed on the abstract system A_Sys k\HS, while our transformation to tock CSP still enables equivalence checks for all considered instances of the abstract and concrete parameterized real-time systems.

Many of the scalability issues stem from the prototypical implementation of our transformations. In future work, we plan to reduce the computational effort of our transformations by making use of parameterized process variables in FDR2 and of data variables in UPPAAL to cope with parameters of process variables. In TCSP2TA, this especially would lead to a reduced location set for transformed processes such that the transformation of parallel processes would be possible for larger models. This is discussed in Section 8.3.

Case Study - A Real-Time Operating System Scheduler

In summary, our transformations can be used to simulate and verify (relatively small) instance models, which already facilitates debugging and instance verification of the parameterized scheduling models in early design phases. However, model and refinement checking can only be employed for the verification of finite models. In particular, we cannot ensure the parameterized system to be correct for all instances. Therefore, it is very important to show the requirements in a comprehensive verification phase as explained in the following. With this, we are able to show that the parameterized system satisfies the given requirements for arbitrary network sizes, `rank` functions, `tt_p` functions, and arbitrary values for `t_init`.

7.3 Comprehensive Verification using Isabelle/HOL

Within the comprehensive verification phase, we first verify that the concrete parameterized scheduling system is weak timed bisimilar to the abstract parameterized scheduling system. Secondly, we verify that the given requirements are satisfied for the abstract parameterized scheduling system. These two steps imply that also the concrete parameterized scheduling system satisfies the given requirements.

7.3.1 Bisimulation Proof

For verifying the semantical equivalence, we define a bisimulation relation consisting of four subrelations.

```
definition BRel :: (( procn , ev ) Process  ×  ( procn , ev ) Process ) set
where   BRel ≡ S1 ∪ S2 ∪ S3 ∪ S4
```

The four subrelations correspond to the following elementary states in which the scheduling system can reside.

7.3 Comprehensive Verification using Isabelle/HOL

- S1 denotes the states where no thread has the control over the computing resources and the schedule list is not empty.

- S2 denotes the states where no thread has the control over the computing resources but the schedule list is empty.

- S3 denotes the states where a thread i has the control over the computing resources.

- S4 denotes the states after the thread who has control has done something or has sent an event.

As an illustrating example, the subrelation S1 is defined as follows.

```
definition S1:: "((procn,ev)Process × (procn,ev)Process) set"
where
"S1 ≡ {(P,Q). ∃ j l init lt lm k r
              sched threads inv wait.
  (* Parameter Conditions *)
       bound 1 k (j#l) ∧ bound 1 k init ∧
       length lt = k ∧ length lm = k ∧
       k>0 ∧
  (* Definition of the Tuple *)
       P = sched |[A]| || k * [alph] threads ∧
       Q = sched |[A]| inv  ∧
  (* Involved Processes *)
       (wait = (WAIT r) ∨ wait = SKIP)
     ∧ (sched = <Sch(j#l,init)>
        ∨ sched = (wait ;; start°j →
                   (yield°j → <Sch(l,init)>
                    □
                    preempt°j → <Sch(insort_key2 rank j l,init)>))
        ∨ sched = (start°j →
                   (yield°j → <Sch(l,init)>
                    □
                    preempt°j → <Sch(insort_key2 rank j l,init)>)))
     ∧ (∀ i. i≥1 ∧ i≤k →
           threads i = <Thread(i,k,lm!(i−1),lt!(i−1))>
         ∨ threads i = asg(Thread (i,k,lm!(i−1),lt!(i−1))))
     ∧ (inv = <Inv(lt,lm,k)> ∨ inv = asg (Inv (lt,lm,k)))}"
```

Case Study - A Real-Time Operating System Scheduler

Each Si consists of process tuples (P,Q) where P denotes the possible states of the concrete parameterized scheduling system while Q denotes the possible states of the abstract parameterized scheduling system. These possibilities are described in the subsequent Involved Processes part. Note that each Si is defined for arbitrary network sizes k>0. The parameter condition bound 1 k list means that all list elements in list are between 1 and k.

The comprehensive bisimulation proof is divided in subproofs where for each tuple of Si it is shown that the respective processes can adequately answer each of their steps and thereby reach states that are again in the bisimulation relation, i.e., are in some subrelation Sj.

```
lemma bisim_CS_AS: "n>0 ⟹ (C_sys n, A_sys n)∈weak_timed_bisimilar"
    apply(coinduct taking: "λ x y. (x,y) ∈ BRel")
...
(* Tuple of S1 *)
apply(erule disjE)
apply(subst (asm) S1_def , rule conjI)
(* ---> *)
...
(* <--- *)
...
(* Tuple of S2 *)
apply(erule disjE)
apply(subst (asm) S2_def , rule conjI)
(* ---> *)
...
(* <--- *)
...
done
```

By splitting the bisimulation in subrelations, the bisimulation proof can be structured in a useful way and the corresponding Isabelle terms are kept small enough to be effectively handled.

7.3 Comprehensive Verification using Isabelle/HOL

7.3.2 Verification of the Timed HML Properties

We now verify that the abstract parameterized scheduling system invariantly satisfies the given requirement formulae R n. To this end, we again define an invariant relation and split it into four subrelations.

```
definition InvRel :: nat ⇒ (procn,ev)Process set
where InvRel k ≡ S1_HS k ∪ S2_HS k ∪ S3_HS k ∪ S4_HS k
```

The four subrelations again denote the elementary states in which the scheduling system can reside. Note that this time, the sets Si_HS are defined with k as an explicit parameter. This is necessary in order to relate the network size with the parameter of the requirements.

As an illustrating example, we present the definition of S1_HS below.

```
definition S1_HS :: nat ⇒ (procn,ev)Process set
where
  S1_HS k ≡ {Q. ∃ j l init lt lm r
              sched inv wait.
    (* Parameter Conditions *)
        sorted_key rank (j#l) ∧ sorted_key rank init ∧
        bound 1 k (j#l) ∧ bound 1 k init ∧
        bound2 lt tt_p ∧
        length lt = k ∧ length lm = k ∧
        ∧ r≥0 ∧ k>0 ∧
    (* Definition of Q *)
        Q = (sched |[A]| inv) \ HS
    (* Involved Processes *)
        (wait = WAIT r ∨ wait = SKIP)
  ∧     (sched = <Sch(j#l,init)>
      ∨ sched = (wait ;; start°j →
                    (yield°j → <Sch(l,init)>
                     □
                     preempt°j → <Sch(insert_key2 rank j l,init)>))
      ∨ sched = (start°j →
                    (yield°j → <Sch(l,init)>
                     □
                     preempt°j → <Sch(insert_key2 rank j l,init)>)))
  ∧ (inv = <Inv(k,lm,lt)> ∨ inv = asg(Inv(k,lm,lt)))}
```

Case Study - A Real-Time Operating System Scheduler

Here, compared to the bisimulation relation, we have the additional parameter condition that the lists j#l and init are sorted according to the rank function. This condition is not needed in the bisimulation relation above but for showing that requirement R2 is satisfied, i.e., that the scheduler respects priorities. For showing requirement R1, we need the additional condition bound2 lt tt_p, which expresses that the residual time of each thread is at most the original worst case execution time as given by tt_p. Note that these conditions do not restrict the actual scheduling system but make valid assertions about the scheduling system explicit.

To show that the set InvRel is indeed an invariant relation, we need to show that each step of a process contained in the invariant again reaches a process in InvRel and that each process contained in the invariant locally satisfies the given requirements.

The first property is verified using the following lemma, where large parts of the bisimulation proof above can be reused.

```
lemma InvRel_step: "⟦ P ∈ InvRel n ; P —<e>→ P' ⟧
                    ⟹ P' ∈ InvRel n"
```

To show that each process of the invariant relation (locally) satisfies the given requirements, we have verified lemmas of the following form (for i replaced accordingly).

```
lemma Si_HS_sat_R: "P ∈ Si_HS n ⟹ P ⊨ R n"
```

This property is quite easy to prove in the case of S2_HS, S3_HS and S4_HS because no initial start event is possible there. However, in the case of S1_HS, this is more complex because we need to ensure that a thread that gets the control over the computing resources eventually yields control or is preempted. This property basically follows from the fact that a thread i takes at most tt_p(i) time units before it *must* yield its control, thereby reaching a state where another thread is given control if the sched-

7.3 Comprehensive Verification using Isabelle/HOL

ule list is not empty. By performing an induction over the remaining time of the thread that has control, we have shown that this property is fulfilled.

Finally, the corresponding proofs concerning the subrelations can be assembled such that we are able to show that InvRel is indeed an invariant relation.

```
lemma AS_sat_R: "n>0 ⟹ (A_sys n) \ HS ∈ invariant (R n)"
  apply(coinduct taking: "λ x. x ∈ InvRel n")
...
(* Process of S1_HS *)
apply(erule disjE)
apply(subst (asm) S1_HS_def)
apply(rule S1_HS_sat_R , simp)
...
(* Process of S2_HS *)
apply(erule disjE)
apply(subst (asm) S2_HS_def)
apply(rule S2_HS_sat_R , simp)
...
apply(rule InvRel_step , auto)
done
```

As described in Section 7.1, the bisimulation lemma and the timed HML lemma together yield the final result that the concrete parameterized scheduling system also satisfies the given requirements. This closes our case study.

```
lemma finish: ∀ n>0. (C_sys n) \ HS ∈ invariant (R n)
```

The comprehensive verification phase has taken us about 35 hours of proof development. Although this is a lot of time, the effort is worth doing because the scheduling system is not only verified for small instances but for arbitrary network sizes, arbitrary priorities and arbitrary maximal computing times of threads. Furthermore, the underlying proofs can be reused for more complex scheduling models. In fact, we started with a non-deterministic scheduler in the first place and were able to reuse about

Case Study - A Real-Time Operating System Scheduler

90 percent of the proofs to verify the prioritized scheduling model. We are confident that again making the scheduling model more complex would enable large parts of the performed proofs to be reused.

7.4 Summary

In this chapter, we have shown that our framework is well-suited for the development and verification of parameterized real-time systems. To this end, we have formalized a simplified but illustrative model of a real-time scheduler. We started with formalizing the Timed CSP model in the Isabelle theorem prover. This has the advantage that its type correctness is guaranteed at all times. Furthermore, our formalization enables us to encode Timed CSP processes in a quite similar way as on a sheet of paper, so that also people being no experts in Isabelle can already take advantage of it. Then, we have shown how our transformation engines are used to verify instances of the parameterized scheduling model. It is currently limited to relatively small models but this already helps a lot when developing and debugging Timed CSP models. Finally, we have verified our scheduler case study comprehensively in the Isabelle/HOL theorem prover. To this end, we have established and verified a bisimulation relation showing the semantical equivalence of the concrete and abstract parameterized scheduling models. Furthermore, we have verified that the abstract parameterized scheduling model satisfies the given requirements. These two steps enable the formal conclusion that also the concrete parameterized scheduling model satisfies the given requirements, which was the main verification goal of this chapter.

With this case study, we have shown the applicability of our verification framework. We have verified timing properties as well as functional properties for a real-time scheduling system. The scheduler copes with ar-

7.4 Summary

bitrarily many threads. Thus, we have verified the scheduling system for all (infinitely many) possible instances. As the comprehensive verification phase in the Isabelle/HOL theorem prover is relatively time-consuming, we could benefit from our transformation engines from Timed CSP to timed automata and tock CSP. The subsequent automatic validation phase helped a lot to debug early versions of our scheduling model.

8 Conclusion and Future Work

In this chapter, we summarize and discuss the results of this thesis. We review the objectives given in the introduction and discuss whether our proposed approach meets them. Then, we give an outlook on future work.

8.1 Conclusion

In this thesis, we have presented a framework that facilitates the mechanical verification of parameterized real-time systems. It is based on a formalization of Timed CSP in the Isabelle/HOL theorem prover. Furthermore, it supports verification using (weak timed) bisimulations and a timed extension of Hennessy-Milner logic (HML). These verification techniques are based on a generic formalization of the notion of timed labeled transition systems (LTSs). Within our formalization we have mechanically proved that the (invariant) satisfaction of timed HML formulae is preserved by weak timed bisimulation. This means that weak timed bisimilar processes satisfy exactly the same timed HML formulae. Furthermore, we have verified that all considered kinds of bisimulation (strong, weak, and weak timed) are observational congruences, which enables compositional verification.

The advantages of our Isabelle/HOL formalization are:

- Possibly infinite (for example, parameterized) real-time systems can be conveniently described using our syntactical representation of Timed CSP processes in Isabelle/HOL.

- Our slightly extended operational semantics of Timed CSP facilitates the observation of hidden events and thus makes internal behavior transparent when needed.

- Different kinds of bisimulation and our timed HML facilitate the mechanical equivalence-based and property-based verification of (timed) LTSs, and in particular of Timed CSP.

- Proofs about Timed CSP processes can be performed with mechanical assistance and are ensured to be correct, i.e., corner cases cannot be overlooked.

- Proofs can be partly automatized using tactics of the Isabelle/HOL theorem prover.

- Our generic formalization of timed LTSs and verification techniques based on them are reusable for other timed process algebras as long as they can be interpreted as timed LTSs.

We have enhanced our framework with capabilities for the automatic verification of finite Timed CSP processes using established verification tools. We have achieved this by adapting and extending the transformation approaches from Timed CSP to timed automata based on [DHQ$^+$08] and from Timed CSP to tock CSP based on [Oua01]. Furthermore, we have implemented our transformation rules such that Timed CSP processes can be transformed automatically.

8.1 Conclusion

The advantages of our transformations are:

- The approaches of [DHQ+08] and [Oua01] are adapted and extended to correctly cope with a large class of (finite) Timed CSP processes respecting the (original) semantics of Timed CSP.

- The UPPAAL model checker can be used for verifying properties of Timed CSP processes.

- The FDR2 refinement checker can be used to verify semantical equivalence of Timed CSP processes (with respect to traces in the tau-priority model).

- Counterexamples are generated if the checks do not succeed in UPPAAL or FDR2. These counterexamples can be used for debugging the original Timed CSP model.

- The UPPAAL simulator can be used for simulating and debugging of Timed CSP processes after translating them to timed automata.

- The ProB and the ProBE simulators can be used for simulating and debugging of Timed CSP processes after translating them to tock CSP. However, τ-events have to be performed in favor of *tock* events manually.

Within our verification framework we have combined the formalization and the automatic verification capabilities in order to define a verification flow for parameterized real-time systems. It consists of the following steps.

1. Formal modeling of a concrete and an abstract parameterized real-time system in our formalization of Timed CSP and formal description of requirements using our formalization of timed HML.

2. Transformation of finite instances of this system for debugging and instance verification.

Conclusion and Future Work

3. Comprehensive, machine-checked and partly automatized verification of the parameterized real-time system in Isabelle/HOL.

In the first step, a designer formally describes a parameterized real-time system using Timed CSP and formalizes it using our Isabelle/HOL formalization of Timed CSP. Furthermore, the requirements for this system are expressed in our timed HML and formalized in Isabelle/HOL. To ease later verification, the designer develops a more abstract version of the parameterized real-time system and also formalizes it in Isabelle/HOL. Thereby, a separation between the verification concerning the parameterization and the verification concerning the properties can be achieved. Altogether, this phase already ensures that the Timed CSP models are type correct and that the (informal) requirements are expressed in a formal language, which makes subsequent verification possible.

In the second step, the designer transforms finite instances of the parameterized real-time systems to timed automata and to tock CSP using our transformation engines. Within UPPAAL, the designer simulates the instance models and verifies them according to the given requirements. If unintended behaviors are present, the designer debugs the original parameterized real-time systems accordingly. Within the FDR2 refinement checker, the designer checks whether instances of the concrete and the abstract parameterized real-time system are equivalent with respect to traces in the tau-priority model. If the corresponding checks fail, the designer is provided with counterexamples, which are used to debug the original parameterized real-time systems. Note that in the phase of debugging, no mechanical proof was yet performed in the Isabelle/HOL theorem prover. Thus, debugging is performed prior to the relatively time-consuming phase of comprehensive verification.

If all checks in UPPAAL and FDR2 are successful, the designer performs the comprehensive verification of the parameterized real-time system using our formalization of the operational semantics of Timed CSP and the definitions concerning bisimulations and our timed HML. To this end, the designer shows that the concrete and abstract parameterized real-time systems are weak timed bisimilar for all possible system parameters. He achieves this goal by providing a bisimulation relation, which contains the concrete and abstract parameterized real-time systems. Furthermore, it has to be shown that for each process pair in the bisimulation relation, each step of the concrete process can be answered by the abstract process and vice versa. When having shown bisimilarity, the designer shows that the abstract parameterized real-time system (invariantly) satisfies the requirements. To achieve this goal, the designer provides an invariant relation, which contains the abstract parameterized real-time system. Furthermore, for each process in the invariant relation it has to be shown that it locally satisfies the requirements and that each derivative is again in the invariant relation. If these two proofs are completed it can be concluded that also the concrete parameterized real-time system satisfies the given requirements.

As presented in Chapter 7, we have successfully applied this verification flow to our case study of a parameterized real-time operating system scheduler. We have checked instances of the scheduling system concerning timing and functional requirements using our transformation engines. Furthermore, we have presented how we performed the bisimulation-based and logic-based verification parts in the comprehensive verification phase.

8.2 Discussion

In Section 1.2, we have introduced a set of objectives for our verification framework. In this section, we discuss how far we have achieved these goals.

Conclusion and Future Work

Coping with General Parameterized Real-Time Systems Other verification approaches for parameterized systems often allow only for restricted parameterized systems to be considered and do not cope with real-time. Especially, real-time systems with a distinguished control process, whose behavior depends on the size of the overall system, and where each process in the network possibly depends on the size of the system, are not supported. Within our framework, we do not have these restrictions as shown, for example, in our case study. In order to be able to omit these restrictions, we had to abandon purely automatic verification. However, with our verification framework, we achieve a trade-off between manual verification and a high degree of automatic support, which enables us to cope with a large class of parameterized real-time systems.

Coping with Real-Time Specifications To cope with real-time specifications, we employ the process calculus of Timed CSP and the notion of weak timed bisimulation. Furthermore, we developed a timed extension of HML where the satisfaction of formulae is preserved by weak timed bisimulation. Timed CSP enables the description of possibly infinite and parameterized real-time systems and provides formal semantics from which we considered the (timed) operational semantics in this thesis. Weak timed bisimulation facilitates the definition of the equivalence of timed processes and provides a powerful proof technique for verifying equivalence of concrete processes. Our timed HML is a minimalistic modal logic with which timing properties and functional properties of real-time processes can be concisely specified and verified.

Comprehensive Machine Assistance To partly automatize proofs concerning the verification of Timed CSP processes and to ensure that these proofs are correct, we employ the Isabelle/HOL theorem prover. Using our formalizations of Timed CSP, (weak timed) bisimulations, and timed

8.2 Discussion

HML, we are able to verify possibly infinite and parameterized real-time systems comprehensively. This means that they are shown to behave correctly for all possible instances of a parameterized real-time systems and all possible values of abstract global parameters. The needed expertise and the required effort for performing mechanical proofs in a theorem prover is comparatively high. For example, performing the proofs concerning our case study took us about 35 hours. Compared to automatic verification, this is quite expensive and for finite systems it would probably not be acceptable. However, for infinite and parameterized systems, automatic verification tools are not applicable and only comprehensive verification can ensure the system to be correct in all cases. Furthermore, the proofs themselves are ensured to be correct. A model checker, for example, could contain implementation errors in optimizations, etc., such that verification results cannot be relied on with full guarantee. In contrast to this, a theorem prover like Isabelle relies only on a very small program core, which needs to be trusted. That bugs in this core, which has been reviewed very often, have been overlooked is very unlikely. In the area of safety-critical systems, there is a need for absolute correct proofs. Our framework facilitates this kind of absolute quality assurance.

Supporting the Development Process with Automatic (Verification) Tools We developed automatic transformations from Timed CSP to timed automata and to tock CSP. Thereby, the UPPAAL tool suite and the FDR2 refinement checker can be employed for the exploration and verification of finite submodels of the overall systems. This additionally gives a designer the possibility to develop an intuition of how the considered models behave and thereby to prepare the subsequent comprehensive verification. Although the set of allowed Timed CSP processes needs to be restricted for a correct verification result, still a large class of processes can be handled by our transformation engines. Factors that currently limit the applicability of our transformations are the number of parallel processes in the system and the range of data variables used in process variables. The number of parallel processes is problematic in the case of the transformation to timed automata because the computation of the syntactical cross product leads to a combinatorial explosion in the number of locations. In both transformations, the range of (data) variables used in process variables is problematic because before transforming a process, all possibly reachable variable valuations are computed such that only processes corresponding to "flat" process variables need to be translated. In the next section, we discuss how we plan to alleviate these problems in future work.

8.3 Future Work

There are a lot of possibilities to apply our framework, to improve it, and to extend it in future work. In the following, we present some of the particularly interesting possible research directions for future work.

Applications In future work, we plan to verify other parameterized real-time systems with our framework. In a student project, a concrete and an abstract parameterized Timed CSP model of the AMBA bus system have been developed. It consists of two distinguished control processes, the arbiter and the decoder. These two components control the access of arbitrarily many slaves and masters to the data bus. The proof obligations for the bisimulation proof have been formalized but have not yet been proved.

In this thesis, we considered the case study of a single-core scheduler. In future work, we would like to elaborate our verification framework for an extended case study of a multi-core scheduler. By this, the abstract parameterized system gets more complex because it needs not only to con-

8.3 Future Work

sider one thread having control over the computing resources but c threads, where c denotes the number of cores.

Transformation Engines As discussed above, our transformation engines currently only work for relatively small Timed CSP processes. To increase the supported size of processes, we plan to improve our transformation engines as follows. Currently, we explicitly compute all reachable process variables before translating them. Thus, a transformation is performed for each single process corresponding to a "flat" process variable. This treatment was easier in order to develop a prototypical implementation of the considered transformations. By using built-in datatypes of UPPAAL and FDR2, we could translate processes corresponding to parameterized process variables as well. For example, in TCSP2TA, a process like $P(m) = a?nm \to P(nm) \square b.m \to P(m)$ could be translated by additionally using a global variable for m in UPPAAL. The use of built-in datatypes in UPPAAL would also make the number of parallel processes, for example within our scheduler case study, less problematic because the number of locations representing a parallel process is drastically reduced using these built-in datatypes. In TCSP2tockCSP, we plan to make use of parameterized process variables of FDR2, as they are directly supported there.

Although a ("paper and pencil") correctness proof for the transformation from Timed CSP to timed automata was given in [DHQ+08], the transformation rules contained flaws as discussed in Section 6.1. Our formalization of the operational semantics gives an ideal basis for performing mechanical correctness proofs for such kinds of transformations. Within a bachelor thesis, we currently formalize a restricted form of intermediate timed automata in Isabelle/HOL in preparation of such a mechanical correctness proof.

As discussed in Section 6.1.3, it is relatively complicated and in general not possible to verify timed HML formulae in the UPPAAL model checker.

Conclusion and Future Work

It would be interesting to examine the possibilities of other timed automata model checkers such as SGM [WH02] or Kronos [DOTY95], which allow for the verification of full Timed CTL. Due to our modular implementation with a language-independent intermediate representation, our transformation engine from Timed CSP to timed automata could easily be changed to target the dialect of another timed automata model checker.

Extending our Timed HML Currently, only finite timed HML formulae, which are combined with coinductive invariants, are supported within our formalization. In future work, it would be advantageous to examine a more general timed HML, which includes least and greatest fix point operators. The main challenge in this research direction is to allow for convenient mechanical proofs to be established for verifying concrete processes. Probably, syntactical proof rules would be necessary to achieve this goal.

Another issue concerning our timed HML is that we would also like to verify properties of the form: whenever some request comes in, it is answered within certain time bounds. This kind of property is only valid in the current version of the logic if no intermediate visible transitions occur between the requesting event and the answering event. In order to verify these properties when also intermediate transitions can occur, we would like to extend our timed HML to allow for ignoring some of the intermediate (visible) events.

Replace Bisimulation by Refinement Weak timed bisimulation is a rather strong notion to relate concrete and abstract processes in a design flow. It especially enforces that two bisimilar processes have exactly the same timing behavior. It would be interesting to replace weak timed bisimulation by a preorder such as proposed in [LV06], which allows for a refined process to evolve faster than the specification process.

8.3 Future Work

Traditionally, denotational semantics are used in the context of CSP based languages. They facilitate the verification on a more abstract semantical basis. Therefore, we could enhance our framework by also integrating the denotational models. The denotational timed failures semantics of Timed CSP, for example, can be characterized operationally [Sch95]. In order to allow for a flexible use of the different semantics, we could formalize this characterization based on our formalization in Isabelle and verify that the process operators have exactly the denotations as usually defined in a denotational semantics mapping. Furthermore, this would enable us to replace weak timed bisimulation by refinement with respect to timed failures.

Within our framework, we propose to develop a more abstract parameterized real-time system in order to separate verification concerns with respect to the parameterization and with respect to timing behavior. As (weak timed) bisimulation is a comparatively strong notion of conformance, this is not always possible because, in general, concurrent timed systems cannot be expressed in terms of a sequential description. It would be very interesting to investigate how well-suited process refinement relations are in order to enable the abstract description of parallel parameterized real-time systems.

Other Process Algebras Finally, we would like to investigate how other timed process algebras can be handled in our verification framework. The process calculus Timed CCS [MT01], for example, could be formalized using its operational semantics in Isabelle/HOL and be interpreted as timed LTS. Then, bisimulations and our timed HML would be inherited and could be explored in the context of this process calculus. Furthermore, our transformation engines could be adapted to support process operators of this language in order to be able to simulate and verify finite (sub)processes.

Conclusion and Future Work

List of Figures

2.1	Labeled Transition System of a Coffee Machine	37
2.2	Timed Labeled Transition System of a Coffee Machine . .	45
2.3	A Simple Timed Automaton	49
2.4	A Producer-Consumer System	52
2.5	Common Structure of an Isabelle-Theory	57
2.6	Proof of Cantor's Theorem in "apply" Notation	65
2.7	Proof of Cantor's Theorem in Isar Notation	66
2.8	Structure of Locales	67
4.1	Conceptual Overview of Our Framework	93
5.1	Excerpt of the Event Step Definition in Isabelle/HOL . . .	128
5.2	Excerpt of the Timed Step Definition in Isabelle/HOL . . .	128

List of Definitions

1	Labeled Transition System	23
2	Timed Labeled Transition System	23
3	Weak Extended Transition Relation	24
4	Weak Timed Extended Transition Relation	25
5	Time-Event Step	25
6	Strong Bisimulation	26
7	Weak Bisimulation	26
8	Weak Timed Bisimulation	27
9	Syntax of Hennessy-Milner Logic	28
10	Semantics of Hennessy-Milner Logic	29
11	Syntax of CSP	31
12	Timed Automaton	48
13	Semantics of a (single) Timed Automaton	49
14	Semantics of a Network of Timed Automata	50
15	Syntax of Timed Hennessy-Milner Logic	104

LIST OF FIGURES

16   Semantics of Timed Hennessy-Milner Logic 105

17   Coinductive Invariant 108

18   Intermediate Representation of Timed Automata 146

Bibliography

[AD94] Rajeev Alur and David L. Dill. A theory of timed automata. *Theoretical Computer Science*, 126:183–235, 1994.

[ADM04] Parosh Aziz Abdulla, Johann Deneux, and Pritha Mahata. Multi-clock timed networks. In *Proceedings of the 19th IEEE Symposium on Logic in Computer Science (LICS)*, pages 345–354. IEEE Computer Society, 2004.

[AJ98] Parosh Aziz Abdulla and Bengt Jonsson. Verifying networks of timed processes (extended abstract). In *Proceedings of the 4th International Conference on Tools and Algorithms for Construction and Analysis of Systems (TACAS)*, pages 298–312. Springer, 1998.

[AK86] Krzysztof R. Apt and Dexter C. Kozen. Limits for automatic verification of finite-state concurrent systems. *Information Processing Letters*, 22:307–309, 1986.

[Bal06] Clemens Ballarin. Interpretation of locales in Isabelle: Theories and proof contexts. In *Proceedings of the 5th International Conference on Mathematical Knowledge Management (MKM)*, pages 31–43. Springer, 2006.

BIBLIOGRAPHY

[Bal10] Clemens Ballarin. Tutorial to locales and locale interpretation. http://www.cl.cam.ac.uk/research/hvg/isabelle/dist/Isabelle2012/doc/locales.pdf, 2010.

[Bar11] Björn Bartels. Verification of distributed embedded real-time systems and their low-level implementations using Timed CSP. In *Proceedings of the 18th Asia-Pacific Software Engineering Conference (APSEC)*, pages 195–202. IEEE Computer Society, 2011.

[BH99] Twan Basten and Jozef Hooman. Process algebra in PVS. In *Proceedings of the International Conference on Tools and Algorithms for the Construction and Analysis of Systems (TACAS)*, pages 270–284. Springer, 1999.

[BJNT00] Ahmed Bouajjani, Bengt Jonsson, Marcus Nilsson, and Tayssir Touili. Regular model checking. In *Proceedings of the 23rd International Conference on Computer Aided Verification (CAV)*, pages 403–418. Springer, 2000.

[BLN03] Dirk Beyer, Claus Lewerentz, and Andreas Noack. Rabbit: A tool for BDD-based verification of real-time systems. In *Proceedings of the 15th International Conference on Computer Aided Verification (CAV)*, pages 122–125. Springer, 2003.

[BLW05] Ahmed Bouajjani, Axel Legay, and Pierre Wolper. Handling liveness properties in (ω-)regular model checking. *Electronic Notes in Theoretical Computer Science*, 138:101–115, 2005.

[BP07] Jesper Bengtson and Joachim Parrow. Formalising the π-calculus using nominal logic. In *Proceedings of the 10th International Conference on Foundations of Software Science and Computational Structures (FoSSaCS)*, pages 63–77. Springer, 2007.

BIBLIOGRAPHY

[Bri93] Lubos Brim. Modal logics in timed process algebras. In *Proceedings of the First North American Process Algebra Workshop (NAPAW)*, pages 13–26. Springer, 1993.

[Bro99] Phillip James Brooke. *A Timed Semantics for a Hierarchical Design Notation*. PhD thesis, University of York, 1999.

[BY04] Johan Bengtsson and Wang Yi. Timed automata: Semantics, algorithms and tools. In *Lectures on Concurrency and Preti Nets*, pages 87–124. Springer, 2004.

[Cam91] Albert John Camilleri. A higher order logic mechanization of the CSP failure-divergence semantics. In *Proceedings of 4th Higher Order Workshop*, pages 123–150. Springer, 1991.

[CK05] Stefano Cattani and Marta Z. Kwiatkowska. A refinement-based process algebra for timed automata. *Formal Aspects of Computing*, 17(2):138–159, 2005.

[Com05] Michael Compton. Embedding a fair CCS in Isabelle/HOL. In *Proceedings of the 18th International Conference on Theorem Proving in Higher Order Logics (TPHOLs)*, Emerging Trends. Springer, 2005.

[Cre01] Sadie Creese. *Data Independent Induction: CSP Model Checking of Arbitrary Sized Networks*. PhD thesis, Oxford University, 2001.

[Dav93] Jim Davies. *Specification and Proof in Real-Time CSP*. Cambridge University Press, New York, NY, USA, 1993.

[DHQ$^+$08] Jin Song Dong, Ping Hao, Shengchao Qin, Jun Sun, and Wang Yi. Timed automata patterns. *IEEE Transactions on Software Engineering*, 34:844–859, 2008.

[DHSZ06] Jin Song Dong, Ping Hao, Jun Sun, and Xian Zhang. A rea-

BIBLIOGRAPHY

soning method for Timed CSP based on constraint solving. In *Proceedings of the International Conference on Formal Engineering Methods (ICFEM)*, Lecture Notes in Computer Science, pages 342–359. Springer, 2006.

[DOTY95] Conrado Daws, Alfredo Olivero, Stavros Tripakis, and Sergio Yovine. The tool KRONOS. In *Proceedings of the DIMACS/SYCON Workshop, Hybrid Systems III: Verification and Control*, pages 208–219. Springer, 1995.

[DS97] Bruno Dutertre and Steve Schneider. Embedding CSP in PVS. An application to authentication protocols. In Elsa Gunter and Amy Felty, editors, *10th International Conference on Theorem Proving in Higher Order Logics (TPHOLs)*, pages 121–136. Springer, 1997.

[EK00] E. Allen Emerson and Vineet Kahlon. Reducing model checking of the many to the few. In *CADE*, pages 236–254. Springer, 2000.

[EN95] E. Allen Emerson and Kedar S. Namjoshi. Reasoning about rings. In *Proceedings of the 22nd ACM SIGPLAN-SIGACT Symposium on Principles of Programming Languages (POPL)*, pages 85–94. ACM, 1995.

[ES00] Neil Evans and Steve Schneider. Analysing time dependent security properties in csp using pvs. In *Proceedings of the 6th European Symposium on Research in Computer Security (ESORICS)*, pages 222–237. Springer, 2000.

[FGT90] William M. Farmer, Joshua D. Guttman, and F. Javier Thayer. IMPS: An interactive mathematical proof system. In *Proceedings of the 10th International Conference on Automated Deduction (CADE)*, pages 653–654. Springer, 1990.

BIBLIOGRAPHY

[FPPZ04]   Yi Fang, Nir Piterman, Amir Pnueli, and Lenore Zuck. Liveness with invisible ranking. In *Proceedings of the 5th International Conference on Verification, Model Checking and Abstract Interpretation (VMCAI)*, pages 223–238. Springer, 2004.

[GHJ07]   Sabine Glesner, Steffen Helke, and Stefan Jähnichen. VATES: Verifying the core of a flying sensor. In *Proceedings of Conquest 2007*. dpunkt Verlag, 2007.

[GL08]   Olga Grinchtein and Martin Leucker. Network invariants for real-time systems. *Formal Aspects of Computing*, 20(6):619–635, 2008.

[GM93]   Michael J. C. Gordon and Thomas F. Melham, editors. *Introduction to HOL: A Theorem Proving Environment for Higher Order Logic*. Cambridge University Press, 1993.

[GRA05]   Michael Goldsmith, Bill Roscoe, and Philip Armstrong. Failures-Divergence Refinement - FDR2 user manual. http://www.fsel.com/fdr2_manual.html, 2005.

[GS92]   Steven M. German and A. Prasad Sistla. Reasoning about systems with many processes. *Journal of the ACM*, 39(3):675–735, 1992.

[HH98]   Jifeng He and C. A. R. Hoare. Unifying theories of programming. In *Proceedings of the 4th International Seminar on Relational Methods in Logic, Algebra and Computer Science (RelMiCS)*, pages 97–99, 1998.

[HM80]   Matthew Hennessy and Robin Milner. On observing nondeterminism and concurrency. In *Proceedings of the 7th Colloquium on Automata, Languages and Programming (ICALP)*, pages 299–309. Springer, 1980.

BIBLIOGRAPHY

[Hoa85] C. A. R. Hoare. *Communicating Sequential Processes*. Prentice Hall International, 1985.

[Int02] Matthias Intemann. Semantische Codierung von Timed CSP in Isabelle/HOL. Diploma thesis, Universität Bremen, 2002.

[IR05] Yoshinao Isobe and Markus Roggenbach. A generic theorem prover of CSP refinement. In *Proceedings of the 11th International Conference on Tools and Algorithms for the Construction and Analysis of Systems (TACAS)*, pages 108–123. Springer, 2005.

[KM95] Robert P. Kurshan and Kenneth L. McMillan. A structural induction theorem for processes. *Information and Computation*, 117(1):1–11, 1995.

[KM07] Eun-Young Kang and Stephan Merz. Predicate diagrams for the verification of real-time systems. *Formal Aspects of Computing*, 19(3):401–413, 2007.

[KMM+97] Yonit Kesten, Oded Maler, Monica Marcus, Amir Pnueli, and Elad Shahar. Symbolic model checking with rich assertional languages. In *Proceedings of 9th International Conference on Computer Aided Verification (CAV)*, pages 424–435. Springer, 1997.

[KWP99] Florian Kammüller, Markus Wenzel, and Lawrence C. Paulson. Locales - a sectioning concept for Isabelle. In *Proceedings of the 12th International Conference on Theorem Proving in Higher Order Logics (TPHOLs)*, pages 149–166. Springer, 1999.

[LA04] Chris Lattner and Vikram Adve. LLVM: A compilation framework for lifelong program analysis & transformation. In *Proceedings of the 2nd IEEE / ACM International Sym-*

BIBLIOGRAPHY

posium on Code Generation and Optimization (CGO), pages 75–85. IEEE Computer Society, 2004.

[Laz99] Ranko Lazić. *A Semantic Study of Data Independence with Applications to Model Checking*. PhD thesis, Oxford University, 1999.

[LF08] Michael Leuschel and Marc Fontaine. Probing the depths of CSP-M: A new FDR-compliant validation tool. In *Proceedings of the 10th International Conference on Formal Engineering Methods (ICFEM)*, pages 278–297. Springer, 2008.

[LL98] François Laroussinie and Kim Guldstrand Larsen. CMC: A tool for compositional model-checking of real-time systems. In *Proceedings of the IFIP Joint Int. Conf. Formal Description Techniques & Protocol Specification, Testing, and Verification (FORTE-PSTV)*, pages 439–456. Kluwer, 1998.

[LLW95] François Laroussinie, Kim Guldstrand Larsen, and Carsten Weise. From timed automata to logic - and back. In *Proceedings of the 20th International Symposium on Mathematical Foundations of Computer Science (MFCS)*, pages 529–539. Springer, 1995.

[LV06] Gerald Lüttgen and Walter Vogler. Bisimulation on speed: A unified approach. *Theoretical Computer Science*, 360(1):209–227, 2006.

[LY93] Kim Guldstrand Larsen and Wang Yi. Time abstracted bisimulation: Implicit specifications and decidability. In *Proceedings of the 9th International Conference on Mathematical Foundations of Programming Semantics (MFPS)*, pages 160–176. Springer, 1993.

[LZDB08] Nikolaos Liveris, Hai Zhou, Robert P. Dick, and Prithviraj

BIBLIOGRAPHY

Banerjee. State space abstraction for parameterized self-stabilizing embedded systems. In *Proceedings of the 7th ACM international Conference on Embedded Software (EM-SOFT)*, pages 11–20. ACM, 2008.

[Mil89] Robin Milner. *Communication and Concurrency*. Prentice-Hall, Inc., Upper Saddle River, NJ, USA, 1989.

[ML09] Tomasz Mazur and Gavin Lowe. Counter abstraction in the CSP/FDR setting. *Electronic Notes in Theoretical Computer Science*, 250(1):171–186, 2009.

[MRH05] Sergio Montenegro, Hans-Peter Röser, and Felix Huber. BOSS: Software and FPGA middleware for the flying-laptop microsatellite. In *Proceedings of the Conference on Data Systems in Aerospace (DASIA)*. ESA Publications, 2005.

[MT90] Faron Moller and Chris M. N. Tofts. A temporal calculus of communicating systems. In *Proceedings of Theories of Concurrency: Unification and Extension (CONCUR)*, pages 401–415. Springer, 1990.

[MT01] Faron Moller and Chris Tofts. TCCS: A temporal calculus of communicating systems (draft). http://www.evernote.com/shard/s1/res/cfe0576b-1689-41b9-afe5-b003549f6022.pdf, 2001.

[Nes92] Monica Nesi. A formalization of the process algebra CCS in higher order logic. Technical report, Cambridge University, 1992.

[Nes99] Monica Nesi. Formalising a value-passing calculus in HOL. *Formal Aspects of Computing*, 11(2):160–199, 1999.

[NPW02] Tobias Nipkow, Lawrence C. Paulson, and Markus Wenzel. *Isabelle/HOL — A Proof Assistant for Higher-Order Logic*, volume 2283 of *LNCS*. Springer, 2002.

[ORS92] Sam Owre, John M. Rushby, and Natarajan Shankar. PVS: a prototype verification system. In Deepak Kapur, editor, *Proceedings of the 11th International Conference on Automated Deduction (CADE)*, volume 607 of *Lecture Notes in Artificial Intelligence*, pages 748–752. Springer, 1992.

[Oua01] Joël Ouaknine. *Discrete Analysis of Continuous Behaviour in Real-Time Concurrent Systems*. PhD thesis, Oxford University Computing Laboratory, 2001.

[OW03] J. Ouaknine and J. Worrell. Timed CSP = closed timed ε-automata. *Nordic Journal of Computing*, 10(2):99–133, 2003.

[PRO03] Process behaviour explorer - ProBE user manual. http://www.fsel.com/probe_manual.html, 2003.

[PRZ01] Amir Pnueli, Sitvanit Ruah, and Lenore D. Zuck. Automatic deductive verification with invisible invariants. In *Proceedings of the 7th International Conference on Tools and Algorithms for the Construction and Analysis of Systems (TACAS)*, pages 82–97. Springer, 2001.

[PXZ02] Amir Pnueli, Jessie Xu, and Lenore D. Zuck. Liveness with (0, 1, ∞)-counter abstraction. In *Proceedings of the 14th International Conference on Computer Aided Verification (CAV)*, pages 107–122. Springer, 2002.

[RE99] Christine Röckl and Javier Esparza. Proof-checking protocols using bisimulations. In *Proceedings of the 10th International Conference on Concurrency Theory (CONCUR)*, pages 525–540. Springer, 1999.

[RH03] Christine Röckl and Daniel Hirschkoff. A fully adequate shallow embedding of the π-calculus in Isabelle/HOL with mechanized syntax analysis. *Journal of Functional Programming*, 13(2):415–451, 2003.

[Ros97] A. W. Roscoe. *The Theory and Practice of Concurrency*. Prentice Hall, Upper Saddle River, NJ, USA, 1997.

[Ros10] A.W. Roscoe. *Understanding Concurrent Systems*. Springer, 2010.

[Sch95] Steve Schneider. An operational semantics for Timed CSP. *Information and Computation*, 116(2):193–213, 1995.

[Sch99] Steve Schneider. *Concurrent and Real Time Systems: The CSP Approach*. John Wiley & Sons, Inc., New York, NY, USA, 1999.

[SLDP09] Jun Sun, Yang Liu, Jin Song Dong, and Jun Pang. PAT: Towards flexible verification under fairness. In *Proceedings of the 21st International Conference on Computer Aided Verification (CAV)*, pages 709–714. Springer, 2009.

[SLR$^+$09] Jun Sun, Yang Liu, Abhik Roychoudhury, Shanshan Liu, and Jin Song Dong. Fair model checking with process counter abstraction. In *Proceedings of the 2nd World Congress on Formal Methods (FM)*, pages 123–139. Springer, 2009.

[Tha95] F. Javier Thayer. An approach to process algebra using IMPS. Technical report, The MITRE Corporation, 1995.

[TW97] Haykal Tej and Burkhart Wolff. A corrected failure divergence model for CSP in Isabelle/HOL. In *Proceedings of the 4th International Symposium of Formal Methods Europe (FME)*, pages 318–337. Springer, 1997.

[Urb08] Christian Urban. Nominal techniques in Isabelle/HOL. *Jour-

BIBLIOGRAPHY

nal of Automated Reasoning, 40(4):327–356, 2008.

[WD96] J. Woodcock and J. Davies. *Using Z—Specification, Refinement, and Proof.* Series in Computer Science. Prentice Hall International, 1996.

[Wen11] Makarius Wenzel. The Isabelle/Isar reference manual. http://www.cl.cam.ac.uk/research/hvg/isabelle/dist/Isabelle2012/doc/isar-ref.pdf, 2011.

[WH02] Farn Wang and Pao-Ann Hsiung. Efficient and user-friendly verification. *IEEE Transactions on Computers*, 51(1):61–83, 2002.

[WH05] Kun Wei and James Heather. Embedding the stable failures model of CSP in PVS. In *Proceedings of the 5th International Conference on Integrated Formal Methods (IFM)*, pages 246–265. Springer, 2005.

[WL89] Pierre Wolper and Vinciane Lovinfosse. Verifying properties of large sets of processes with network invariants. In *Proceedings of the International Workshop on Automatic Verification Methods for Finite State Systems*, pages 68–80. Springer, 1989.

[WWB10] Kun Wei, Jim Woodcock, and Alan Burns. Embedding the timed Circus in PVS. Technical report, University of York, 2010.

[WWH05] Farn Wang, Rong-Shiung Wu, and Geng-Dian Huang. Verifying timed and linear hybrid rule-systems with RED. In *Proceedings of the 17th International Conference on Software Engineering and Knowledge Engineering (SEKE)*, pages 448–454, 2005.

Publications by Thomas Göthel

[BGG10] Björn Bartels, Sabine Glesner, and Thomas Göthel. Model transformations to mitigate the semantic gap in embedded systems verification. In *International Colloquium on Graph and Model Transformation – on the occasion of the 65th birthday of Hartmut Ehrig*, 2010. http://journal.ub.tu-berlin.de/eceasst/article/view/418/401.

[GBGK10] Sabine Glesner, Björn Bartels, Thomas Göthel, and Moritz Kleine. The VATES-diamond as a verifier's best friend. In Simon Siegler and Nathan Wasser, editors, *Verification, Induction, Termination Analysis*, volume 6463 of *Lecture Notes in Computer Science*, pages 81–101. Springer, 2010. http://www.springerlink.com/content/x4717236n15rpv23/fulltext.pdf.

[GG09] Thomas Göthel and Sabine Glesner. Machine-checkable Timed CSP. In *Proc. of The First NASA Formal Methods Symposium*, pages 126–135. NASA Conference Publication, 2009. http://ti.arc.nasa.gov/m/events/nfm09/proceedings.pdf.

[GG10a] Thomas Göthel and Sabine Glesner. An approach for machine-assisted verification of TimedCSP specifications. *Innovations in Systems and Software Engineering - A NASA Journal*, 7, 2010. http://www.springerlink.com/content/g3816687x45322m0/fulltext.pdf.

[GG10b] Thomas Göthel and Sabine Glesner. Towards the semi-automatic verification of parameterizedreal-time systems using network invariants. In *Proceedings of the 8th IEEE International Conference on Software Engi-*

Publications by Thomas Göthel

neering and Formal Methods. IEEE Computer Society, 2010. http://ieeexplore.ieee.org/stamp/stamp.jsp?tp=&arnumber=5637374.

[Göt07] Thomas Göthel. Formalisierung von Timed CSP im Theorembeweiser Isabelle/HOL. Diploma thesis, Technische Universität Berlin, 2007.

[KBG+11] Moritz Kleine, Björn Bartels, Thomas Göthel, Steffen Helke, and Dirk Prenzel. LLVM2CSP: Extracting CSP models from concurrent programs. In M. Bobaru, K. Havelund, G. Holzmann, and R. Joshi, editors, *3rd NASA Formal Methods Symposium*, number 6617, pages 500 – 505. Springer, 2011. http://www.springerlink.com/content/764xw43j1h056xu2/fulltext.pdf.

[KBGG09] Moritz Kleine, Björn Bartels, Thomas Göthel, and Sabine Glesner. Verifying the implementation of an operating system scheduler. In *3rd IEEE International Symposium on Theoretical Aspects of Software Engineering (TASE '09)*, pages 285–286. IEEE Computer Society, 2009. http://ieeexplore.ieee.org/stamp/stamp.jsp?arnumber=05198513.

[KG10] Moritz Kleine and Thomas Göthel. Specification, verification and implementation of business processes using CSP. In *4th IEEE International Symposium on Theoretical Aspects of Software Engineering*, pages 145–154. IEEE Computer Society, 2010. http://ieeexplore.ieee.org/stamp/stamp.jsp?tp=&arnumber=5587716.

Diploma and Master Theses Supervised by Thomas Göthel

[Lah09] Michael Lahs. Spezifikation und Analyse von Kernkomponenten und Anwendungen eines Echtzeitbetriebssystems mit TCOZ. Diploma thesis, Technische Universität Berlin, 2009.

[Wu10] Hongyun Wu. Automatische Transformation von Timed CSP nach Timed Automata. Master's thesis, Technische Universität Berlin, 2010.

[Zho10] Caifeng Zhou. Automatische Transformation von Timed CSP nach Tock CSP. Master's thesis, Technische Universität Berlin, 2010.

i want morebooks!

Buy your books fast and straightforward online - at one of world's fastest growing online book stores! Environmentally sound due to Print-on-Demand technologies.

Buy your books online at
www.get-morebooks.com

Kaufen Sie Ihre Bücher schnell und unkompliziert online – auf einer der am schnellsten wachsenden Buchhandelsplattformen weltweit! Dank Print-On-Demand umwelt- und ressourcenschonend produziert.

Bücher schneller online kaufen
www.morebooks.de

 VDM Verlagsservicegesellschaft mbH
Heinrich-Böcking-Str. 6-8 Telefon: +49 681 3720 174 info@vdm-vsg.de
D - 66121 Saarbrücken Telefax: +49 681 3720 1749 www.vdm-vsg.de

Printed by Books on Demand GmbH, Norderstedt / Germany